Anatomy:
Review for New National Boards

Anatomy:
Review for New National Boards

Kurt E. Johnson, Ph.D.
Professor of Anatomy
George Washington University Medical Center
Washington, D.C.

Frank J. Slaby, Ph.D.
Associate Professor of Anatomy
George Washington University Medical Center
Washington, D.C.

J&S Publishing Company, Alexandria, Virginia

Composition and Layout: Ronald C. Bohn, Ph.D.
Cover Design: Kurt E. Johnson, Ph.D.
Printing Supervisor: Anthony Spagnolo
Printing: Deadline Press, Washington, D.C.

Library of Congress Catalog Card Number 92-81332

ISBN 0-9632873-03

Dedication

Kurt E. Johnson would like to dedicate his portion of this book to the maternal-fetal unit of Julie M. Okkema, M.D. and Justine Victoria Johnson. Frank J. Slaby would like to dedicate his portion of this book to his wife Susan K. McCune, M.D.

Table of Contents

Preface

This book is designed to enable you to review in just 1-2 days all of the basic anatomical sciences you studied in the first year of medical school: Cell Biology, Histology, Embryology, Gross Anatomy and Neuroanatomy. We have been able to condense a review of all basic anatomical sciences into a single book because of the new format of the National Board Part I Exam. It is no longer prudent to review exhaustively the basic science courses because the new examination format no longer rewards an encyclopedic knowledge of the basic sciences. Instead, the new exams test knowledge of the scientific basis of disease and injury and the ability to apply basic scientific information to the clinical reasoning process. Consequently, the most efficient way to study for the new exam is 1) to review only the most clinically relevant material from each basic science course and 2) to focus on the application of this material to the solution of clinical problems. These two new study features form the core of this text.

If you answer every question and read all the tutorials in this book, you can cover within 2 days all of the most clinically relevant information from your basic anatomical science courses. You will find that many anatomical facts reviewed or learned anew will be presented in the context of a clinical case or an illustration. We hope that the clinical cases and illustrations will enhance your understanding and recall of the information. Finally, you will learn from the tutorials how anatomical information is used by knowledgeable physicians to understand the courses of diseases, the mechanisms of injuries and the significance of abnormal findings.

Kurt E. Johnson, Ph.D.
Frank J. Slaby, Ph.D.
Washington, D.C.
April, 1992

Acknowledgements

The authors would like to thank Ronald C. Bohn, Ph. D., Associate Professor, Department of Anatomy, The George Washington University Medical Center for writing Chapter IV and for his assistance in formatting the final documents for publication. We would also like to thank Mark R. Adelman, Ph. D., Associate Professor, Department of Anatomy, Uniformed Services University of the Health Sciences for reading and correcting Chapters I and II. We would like to acknowledge the use of the excellent illustrations by Diane Abeloff, A.M.I. published in Medical Art: Graphics for Use, Williams and Wilkins, Baltimore, 1982. Mr. Fred G. Lightfoot, Electron Microscopy Suite Supervisor, George Washington University Medical Center kindly supplied the scanning electron micrograph of fractured renal cortex and we are grateful for his help. We would also like to thank Raymond J. Walsh, Professor and Chairman, Department of Anatomy, George Washington University Medical Center for his support of this project.

Disclaimer

The clinical information presented in this book is accurate for the purposes of review for licensure examinations but in no way should be used to treat patients or substituted for modern clinical training. Proper diagnosis and treatment of patients requires comprehensive evaluation of all symptoms, careful monitoring for adverse responses to treatment and assessment of the long-term consequences of therapeutic intervention.

CHAPTER I

CELL BIOLOGY AND HISTOLOGY

Items 1-5

For each statement of structure, function or embryological origin, select the most appropriate structure in the photomicrograph.

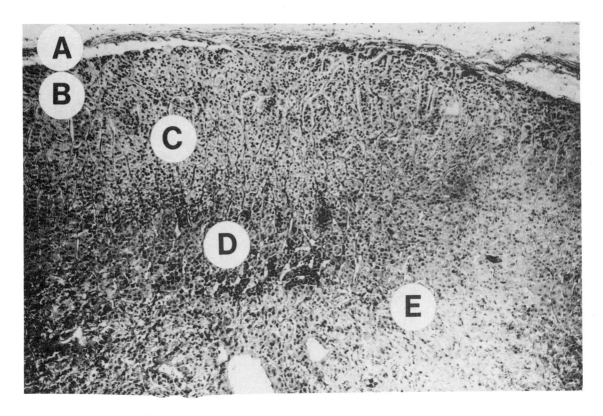

1. This structure is derived from the neural crest.

2. This structure has abundant smooth endoplasmic reticulum, mitochondria with tubular cristae and an abundance of large lipid droplets.

3. This structure secretes mineralocorticoids.

4. This structure secretes sex steroids and is derived from the lining of the primitive coelomic cavity.

5. Epithelial cells in this structure are arranged in long straight cords running parallel to long straight sinusoids.

ANSWERS AND TUTORIAL ON ITEMS 1-5

The answers are: **1-E, 2-C, 3-B, 4-D, 5-C.** This is a photomicrograph of the human adrenal gland. It is surrounded by a thin connective tissue capsule (A). The adrenal cortex (B-D) is derived from proliferation of the mesodermally derived coelomic epithelium and consists of three layers of cells. The outer layer is the zona glomerulosa (B). It is a source of mineralocorticoids. The middle and thickest layer is the zona fasciculata (C). It consists of long, straight epithelial cords of epithelial cells arranged between long sinusoids. The cortical cells all have an abundance of smooth endoplasmic reticulum and mitochondria with tubular cristae, ultrastructural features common to all steroid secreting cells. The cells of the zona fasciculata have a foamy appearance in the light microscope due to a profusion of large lipid rich vacuoles. They secrete glucocorticoids. The cells of the zona reticularis (D) secrete sex steroids. The adrenal medulla (E) is derived from the neural crest and consists of two populations of secretory cells, one secreting epinephrine and the other secreting norepinephrine.

Items 6-8

    (A) Lactotropes
    (B) Gonadotropes
    (C) Corticotropes
    (D) Thyrotropes
    (E) Somatotropes

6. These acidophils are found in the pars distalis. When they are hyperactive in adults, acromegaly results.

7. These basophils secrete luteinizing hormone and follicle-stimulating hormone.

8. Adrenalectomy would result in degranulation of these cells due to release of ACTH.

The answers are: **6-E, 7-B, 8-C**. All of these cell types are found in the adenohypophysis of the pituitary gland. Corticotropes (large granules), gonadotropes (medium granules) and thyrotropes (small granules) are all basophils. Lactotropes and somatotropes are acidophils. Corticotropes secrete ACTH which in turn stimulates steroid secretion from the adrenal cortex. Adrenalectomy would remove the feedback inhibition of ACTH secretion and thus cause massive degranulation of corticotropes. Gonadotropes secrete FSH and LH (ICSH), hormones that are involved in the regulation of gamete formation in the gonads. Thyrotropes secrete TSH which stimulates thyroxine secretion in the thyroid gland. Lactotropes produce prolactin which stimulates development of the mammary glands and lactation. Somatotropes secrete growth hormone. Excessive secretion of growth hormone in adults leads to acromegaly.

Items 9-11

Examine the high power light micrograph of a mature ovarian follicle below and then choose the most appropriate labeled structure to match the functional role or morphological description of this structure.

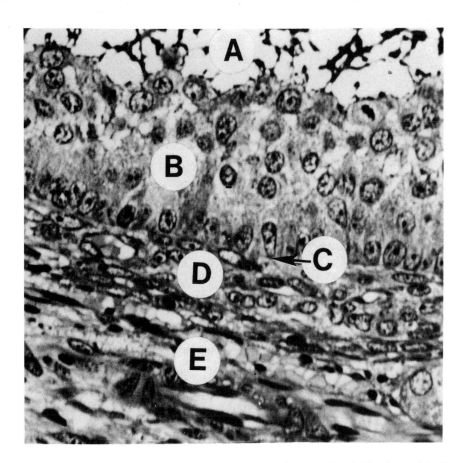

9. This structure is rich in laminin and is the outer boundary of the follicular epithelium.

10. These cells secrete androstenedione, an estradiol precursor.

11. These cells secrete liquor folliculi and the zona pellucida before ovulation and steroids after ovulation.

The answers are: **9-C, 10-D, 11-B.** This is a high magnification light micrograph of a portion of a mature ovarian follicle and the adjacent thecal and stromal cells associated with this follicle. The follicle is bounded by a basement membrane (C). Inside the basement membrane, there are several layers of granulosa cells (B). Not shown in this photograph is the oocyte surrounded by a group of granulosa cells known collectively as the cumulus oophorus. The granulosa cells are thought to be involved in supporting the development of the follicle in several ways. First, they proliferate and contribute to the growth of the follicle. Second, they secrete a viscous liquor folliculi (A) into a growing follicular antrum. Third, they secrete estradiol synthesized from androstenedione derived from the cells of the theca interna (D). After ovulation, the follicle is converted into a steroid secreting corpus luteum. The granulosa cells differentiate into granulosa lutein cells of the corpus luteum. The theca interna cells also contribute to the formation of the corpus luteum by differentiating into theca lutein cells. The steroids secreted by the developing follicle prior to ovulation stimulate the growth of endometrial glands. After ovulation, hormones of the corpus luteum stimulate the secretory activity of endometrial glands in preparation for implantation of the conceptus in the wall of the uterus.

Items 12-14

You encounter a patient who is phenotypically female. Karyotype analysis reveals a normal looking 46, XY karyotype. Molecular biological investigation of the Y chromosome DNA sequences shows a deletion of the testis determining factor (TDF).

12. The testis determining factor encodes for a

    (A) zinc finger protein
    (B) ras oncogene
    (C) fos oncogene
    (D) insulin-like growth factor
    (E) zipper protein

13. If the patient had a 46,XX karyotype with the TDF region of the Y chromosome translocated into the X chromosome the patient would be

    (A) phenotypically female
    (B) phenotypically male
    (C) a hermaphrodite
    (D) a Turner syndrome
    (E) a Klinefelter syndrome

14. The testis determining factor gene product is a

    (A) translation factor
    (B) repressor
    (C) DNA polymerase inhibitor
    (D) transcription factor
    (E) reverse transcriptase

ANSWERS AND TUTORIAL ON ITEMS 12-14

The answers are: **12-A, 13-B, 14-D.** The testis determining factor is a gene product sharing significant homology with other zinc finger proteins, a class of transcription factors. Growth factors and oncogene products are also transcription factors. When the DNA sequence encoding the testis determining factor is deleted from the Y chromosome, sex reversal results. A male karyotype and a female phenotype will occur. Conversely, when the DNA sequence encoding for the testis determining factor is translocated from the Y chromosome to the X chromosome, sex reversal will result again. In this instance, a female karyotype and a male phenotype will be associated. The testis determining factor drives the differentiation of the indifferent gonad in the male direction. Differentiation of a testis then directs sexual differentiation in a male direction. Turner syndrome is caused by a monosomy of the X chromosome (45, X). Klinefelter syndrome is cause by aneuploidy of sex chromosomes (47, XXY).

Items 15-17

Recent discoveries in developmental genetics reveal a connection between ß transforming growth factor (ß-TGF), nerve growth factor (NGF), homeodomain proteins, and the fos gene product.

15. All of these polypeptides

      (A) are DNA polymerases
      (B) are translation factors
      (C) are transcription factors
      (D) bind to mRNA
      (E) block tRNA turnover

16. Their chief mechanism of action is to

      (A) promote specific mRNA synthesis
      (B) inhibit DNA synthesis
      (C) promote specific tRNA synthesis
      (D) suppress gene expression
      (E) activate genes for rRNA synthesis

17. They are thought to result in differentiation by

      (A) repression of gene expression
      (B) controlling differential gene expression
      (C) promoting turnover of the rough endoplasmic reticulum
      (D) converting rough endoplasmic reticulum into smooth endoplasmic reticulum
      (E) stimulation of mitochondrial DNA polymerase

The answers are: **15-C, 16-A, 17-B.** ß-TGF, NGF, homeodomain proteins, and the fos gene product are all examples of transcription factors. They bind to DNA causing its uncoiling and promote its transcription into specific mRNAs. The action of these transcription factors is thought to represent the fundamental basis for differential gene expression. All cells in an organism are thought to have the same DNA complement. The formation of the multiple differentiated cell types within an organism, e.g. neurons, muscle cells and liver cells, is due to selective activation of some subset of genes within the precursor of that individual cell type. Once these genes are activated by transcription factors, specific transcription and translation results in the formation of an array of proteins peculiar to that highly differentiated cell type.

Items 18-20

Experimental vascular perfusion of the testis with lanthanum nitrate (an electron dense, low molecular weight tracer) is followed by fixation of seminiferous tubules. Subsequently, seminiferous tubules are prepared for transmission electron microscopy.

18. Lanthanum nitrate would be found in all of the following anatomical locations **EXCEPT**:

    (A) interstitial tissue
    (B) surrounding Leydig cells
    (C) around primordial germ cells
    (D) around immature primary spermatocytes
    (E) around spermatids

19. What anatomical structural arrangement prevents penetration of lanthanum nitrate into the adluminal compartment?

    (A) Desmosomes
    (B) Zonula adherens
    (C) Tight junctions
    (D) Gap junctions
    (E) The basement membrane of the seminiferous epithelium

20. These structural elements also function in the formation of all of the following morphological barriers **EXCEPT**:

    (A) blood-urine barrier in kidney
    (B) blood-brain barrier
    (C) blood-bile barrier
    (D) impermeable continuous capillaries
    (E) barrier preventing leakage of digestive enzymes from intestinal lumen

ANSWERS AND TUTORIAL ON ITEMS 18-20

The answers are: **18-E, 19-C, 20-A**. Tight junctions are an essential feature of many epithelial layers. Epithelia line cavities and cover surfaces. They have tight lateral junctions that allow them to serve as boundary tissues, separating one compartment in the body from another. For example, intestinal epithelial cells are joined together by apical junctional complexes. The junctional complex consists of an apical zonula occludens or tight junction, a zonula adherens just deep to the zonula occludens and a macula adherens (desmosome) deep to the zonula adherens. The tight junction is a region of fusion of the outer leaflets of the plasma membranes of adjacent cells. It provides a hydrophobic barrier preventing the contents of the intestinal lumen (digestive enzymes) from diffusing into the lateral spaces between cells. Tight junctions are also present between capillary endothelial cells in continuous capillaries where they serve as the anatomical basis for the blood-brain barrier and between liver parenchymal cells where they serve as the anatomical basis for the blood-bile barrier. The blood-urine barrier in the kidney is more complex. The glomerular basement membrane and the filtration slit diaphragms between podocyte foot processes serve as the anatomical barrier between blood and urine. In the seminiferous epithelium, Sertoli cells form a continuous epithelial layer. The spermatogenic cell line is lodged in the spaces between Sertoli cells. A complex web of tight junctions between adjacent Sertoli cells divides the seminiferous epithelium into a basal compartment containing only spermatogonia, preleptotene primary spermatocytes and leptotene primary spermatocytes. Later stages in spermatogenesis including late primary spermatocytes, secondary spermatocytes, spermatids and spermatozoa are contained within the adluminal compartment apical to the web of tight junctions. This elaborate system of tight junctions prevents exposure of the immune system to foreign antigens of mature gametes. During fetal development, the male gonad becomes equipped with spermatogonia prior to the time in development when the immune system gains the ability to discriminate between native and foreign antigens. Thus, spermatogonia are recognized as native antigens. Spermatogenesis does not begin until puberty, long after the establishment of the immunological sense of native and foreign antigens. Consequently, the surface antigens peculiar to spermatozoa would be recognized as foreign antigens were it not for the tight junctions excluding these foreign antigens from immune surveillance. Other mechanisms also ensure that there is no contact between seminal antigens and the male circulatory system.

Examine the labeled photomicrograph of a developing bone of the appendicular skeleton below. Match the lettered structure in the photomicrograph with the most appropriate description of its developmental fate or role in endochondral bone formation in the questions.

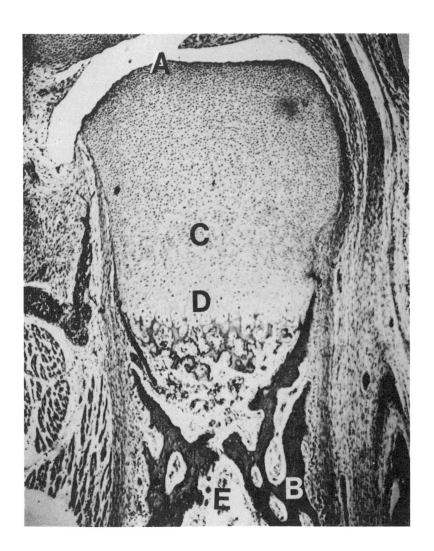

21. This structure represents a current site of osteoid deposition.

22. This region contains many chondrocytes actively involved in mitosis.

23. This region contains hypertrophic chondrocytes.

24. This region persists as hyaline cartilage even in the adult bone of the appendicular skeleton.

ANSWERS AND TUTORIAL ON ITEMS 21-24

The answers are: **21-B, 22-C, 23-D, 24-A**. All bones of the appendicular skeleton as well as vertebrae, the base of the skull, ribs and the sternum are formed by the ossification of cartilaginous models of these bones. This type of bone formation is called endochondral bone formation. The bones of the vault of the skull and certain facial bones are formed spontaneously in mesenchyme without a pre-existing cartilaginous model by intramembranous bone formation. During endochondral bone formation, the cartilaginous model grows by mitotic proliferation of chondrocytes in the zone of proliferation (C) or by addition of new chondrocytes to the outer surface of the growing cartilaginous model. Certain mesenchymal cells in the future metaphysis of the bone differentiate into osteoblasts. These cells secrete an extracellular matrix characteristic of bone which calcifies rapidly, forming a bony collar (B) around the cartilage. The nutrient supply of chondrocytes is restricted by the bony collar and they therefore undergo hypertrophy (D) and then degenerate and die, leading to the formation of a marrow cavity (E). Meanwhile, growth continues in the zone of proliferation. New cartilage is rapidly converted into bone. Once the final size of the bone has been established at puberty, further proliferation of the cartilage is not possible. The hyaline cartilage on the articular surface (A) of these bony models persists as the articular cartilage.

Items 25-27

    (A)   Zonula occludens
    (B)   Zonular adherens
    (C)   Macula adherens
    (D)   Gap junction
    (E)   Microfilaments

For each component of the junction complex, select the most appropriate associated functional role of this component.

25. This is a location where aqueous channels between cells assure free passage of small molecules.

26. This structure is thought to be involved in intercellular adhesions. Here one finds dense plaques on the cytoplasmic face of apposed membranes which serve as insertion sites for tonofilaments.

27. This is a site of fusion of outer leaflets of the plasma membrane. It is a tight junction preventing luminal materials from leaving the lumen.

The answers are: **25-D, 26-C, 27-A.** Epithelial tissue defines boundaries and establishes compartments in the human body. For example, the lumen of the small intestine contains a complicated mixture of digestive enzymes capable of digesting the wall of the small intestine. The contents of the lumen of the GI tract are isolated from the sensitive wall of the gut by a membrane specialization known as the junctional complex. At the most apical portion of the junctional complex there is a tight junction where the outer leaflets of the membranes fuse into an occluding junction called the zonula occludens. This structure extends belt-like around the apex of the columnar epithelial cells and makes a seal between the lumen and the lateral extracellular fluid environment. In freeze-fracture-etch, the zonula occludens sometimes occurs as an anastomosing network of ridges (points of membrane fusion) representing multiple barriers to movement of molecules from the lumen to the lateral extracellular compartment. Below the zonula occludens there is a divergence of the plasma membrane with a clear separation of 10 to 15 nm. This structure is called the zonula adherens. There is simple membrane apposition with variable amounts of electron dense material in the intervening 10-15 nm gap. Numerous 6 nm microfilaments radiate away from the zonula adherens into the cytoplasmic matrix of apposed cells. This structure is usually described as an adhesive junction. The macula adherens (desmosome) is found below the zonula adherens. At the macula adherens, the plasma membranes diverge to 25-30 nm. There is an intermediate dense line running between the cells. On the inner face of each apposed plasma membrane there is a plaque of electron dense material. Long bundles of 10 nm intermediate filaments called tonofilaments radiate away from the plaque of electron dense material. The macula adherens is thought to be a structure holding cells together. Gap junctions are also commonly associated with the junctional complex. In the gap junction (sometimes called a nexus), the outer leaflets of the membranes of adjacent cells approach to within 2 nm, but a small but definite gap remains. Gap junctions are composed of hexagonal arrays of barrel-shaped structures with six subunits arranged around an electron lucid central core. This core is an aqueous channel between closely apposed cells, allowing the free passage of ions and other small molecules between epithelial cells.

Items 28-31

Examine the transmission electron micrograph below and then choose the correct answer.

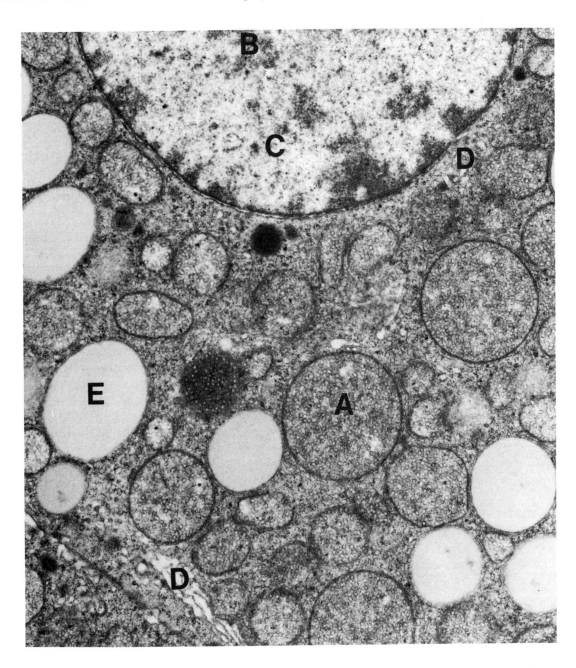

28. Which physiological function is most characteristic of a cell with this ultrastructure?

    (A) Mucus secretion
    (B) Zymogen granule production
    (C) Steroid synthesis
    (D) Glycogen storage
    (E) Motility

29. This kind of a cell would be most prominent in which anatomical location?

   (A) Gastric mucosa
   (B) Pancreatic acini
   (C) Adrenal cortex
   (D) Liver parenchyma
   (E) Wall of the urinary bladder

For each physiological function, select the corresponding labeled structure in the transmission electron micrograph.

30. Site of storage of cholesterol esters, precursors for the steroid secretion products of this cell.

31. Produces ATP and is involved in steroid biosynthesis.

ANSWERS AND TUTORIAL ON ITEMS 28-31

The answers are: **28-C, 29-C, 30-E, 31-A**. This is a transmission electron micrograph of a cell from the adrenal cortex. It has an ultrastructure characteristic of cells secreting steroids including an abundance of smooth endoplasmic reticulum (D) (not well illustrated here), numerous lipid droplets (E), large round mitochondria with tubulovesicular cristae (A) and small dense bodies. The nucleolus (B) in a nucleus (C) is prominent. This cell synthesizes cortisol from cholesterol esters stored in lipid droplets (E). Cholesterol is released from the lipid droplets and enters mitochondria where it is converted into pregnenolone. In the smooth endoplasmic reticulum, pregnenolone is converted to progesterone and then 17-deoxycorticosterone, which is finally converted into cortisol by mitochondrial enzymes. The factors controlling the shuttling of different intermediates from lipid droplets to mitochondria to smooth endoplasmic reticulum and back to mitochondria are not well understood. Also, the mechanism of steroid secretion is controversial with most authors favoring direct release of steroids by diffusion. Steroid secreting cells are abundant in the adrenal cortex, in the corpus luteum, in the placenta and in the interstitium of the testis (Leydig cells).

    (A) Lumen of intestine
    (B) Microvilli
    (C) Smooth endoplasmic reticulum of intestinal absorptive epithelial cell
    (D) Golgi apparatus of intestinal absorptive epithelial cell
    (E) Lacteals

Following a lipid rich meal, dietary lipids are processed and transported to the liver. Choose the most appropriate anatomical location for the physiological process described in the questions below.

32. Chylomicra are synthesized from triglycerides, glycolipids and proteins.

33. Free fatty acids and monoglycerides are converted into triglycerides.

34. Chylomicra are transported via these lymphatic vessels to the systemic circulation.

ANSWERS AND TUTORIAL ON ITEMS 32-34

The answers are: **32-D, 33-C, 34-E.** Dietary fat, composed mainly of triglycerides, is hydrolyzed to fatty acids and monoglycerides in the lumen of the small intestine by the action of lipases secreted by the pancreas. Fatty acids and monoglycerides diffuse across the plasma membranes of intestinal microvilli and into the cisternae of the smooth endoplasmic reticulum located in the apical cytoplasm of absorptive cells. Here, the fatty acids and monoglycerides are resynthesized into triglycerides. Next, the triglycerides are transported to the Golgi apparatus where they are further processed by addition of glycolipids and proteins to form the chylomicra. Chylomicra are now expelled from the lateral borders of absorptive epithelial cells into intercellular spaces where they move across the epithelial basement membrane and finally enter the lumen of blind ending lymphatic capillaries (lacteals) in the core of intestinal villi. Lymphatic vessels conduct the chylomicra to the systemic circulation. The contraction of slips of smooth muscle in the villi probably aids in proximal transport of chylomicra.

Examine the labeled photomicrograph below and then match the lettered structure with the most appropriate description of its microscopic anatomy, developmental origins or physiological role.

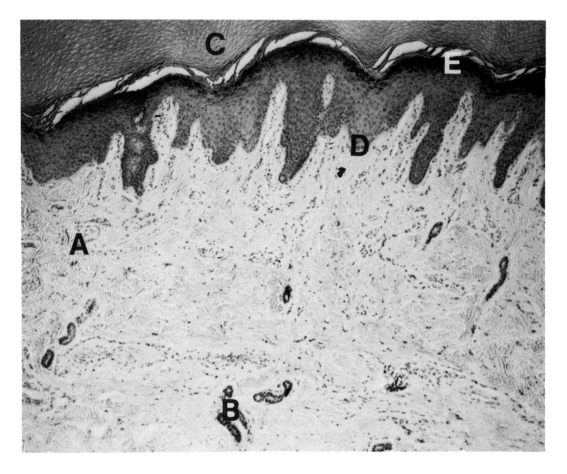

35. This structure is derived from segmentally-arranged dermomyotomes which are formed by proliferation in derivatives of the intraembryonic mesoderm.

36. This structure contains cells with nuclei protected on their apical side by a layer of melanin granules.

37. This structure contains undifferentiated stem cells that are mitotically active.

38. This structure functions to prevent desiccation of the body from within and contains cross-linked keratin forming a hydrophobic barrier.

39. This structure is formed by invagination of ectodermal derivatives.

40. This structure contains keratohyalin granules.

ANSWERS AND TUTORIAL ON ITEMS 35-40

The answers are: **35-A, 36-D, 37-D, 38-C, 39-B, 40-E.** This is a photomicrograph of human skin. It consists of an epithelial epidermis and a connective tissue dermis (A). The dermis is derived from a component of the somite known as the dermomyotome. Somites are segmentally arranged structures forming in the intraembryonic mesoderm. The most basal layer of the epidermis is the stratum germinativum (D). It consists of undifferentiated stem cells capable of repeated rounds of mitosis to produce all of the more apical epidermal layers. Because these stems cells are mitotically active and exposed to ionizing mutagenic UV irradiation from the sun, their nuclei are capped with melanin granules that absorb UV rays. The stratum granulosum (E) is intermediate between the deep stratum germinativum and the superficial stratum corneum (C). The stratum granulosum contains basophilic keratohyalin granules that contribute components to the keratinization process. The stratum corneum consists of many layers of dry, dead squamous cells. Each cell contains a high concentration of keratin, a highly cross-linked protein made up of many hydrophobic amino acids. This layer prevents unwanted substances from entering the body and also prevents loss of interstitial water. An eccrine sweat gland (B) is also shown. It is formed by invagination of the surface ectoderm deep into the dermis. Surface ectoderm eventually differentiates into the epidermis and all epidermal appendages including sweat glands, sebaceous glands, hair, eyelashes and nails under the inductive influence of underlying dermal connective tissues.

Examine the scanning electron micrographs below. The area in the box in the left micrograph is shown at higher magnification in the right micrograph. Match the structure in the micrograph with the most appropriate description of its microscopic anatomy or physiological role.

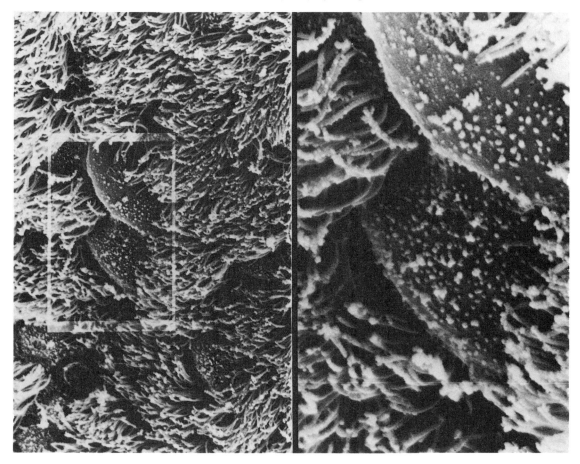

41. The best functional description of the two cells inside the box is

    (A) their secretion product has a digestive function
    (B) their secretion product traps inspired debris
    (C) their secretion product moistens ingested food
    (D) they are involved in absorption of materials from the lumen of this organ
    (E) they are involved in excretion of urea

42. The best microanatomical description of the predominant cell type in this micrograph is

    (A) abundant smooth endoplasmic reticulum and mitochondria with tubular cristae
    (B) abundant rough endoplasmic reticulum and many zymogen granules
    (C) basal bodies and apical mitochondria
    (D) basal infoldings and basal mitochondria
    (E) apical microvilli

43. The most abundant and physiologically significant cytoskeletal element in the predominant cell type is

   (A) actin-containing microfilaments
   (B) desmin-containing intermediate filaments
   (C) keratin-containing intermediate filaments
   (D) vimentin-containing intermediate filaments
   (E) dynein-containing microtubular pairs

44. Both kinds of cells would be found in all of the following anatomical locations EXCEPT:

   (A) uterus
   (B) uterine tubes
   (C) trachea
   (D) bronchi
   (E) terminal bronchioles

ANSWERS AND TUTORIAL ON ITEMS 41-44

The answers are: **41-B, 42-C, 43-E, 44-E.** This is a scanning electron micrograph of the apical surface of the human tracheal epithelium. It consists mainly of abundant ciliated cells (the predominant cell type in the photomicrograph) and sparse goblet cells (in the box). Goblet cells secrete a thick mucus that traps inspired debris in the respiratory system. Ciliated cells have many cilia consisting of nine peripheral doublets of microtubules and a central pair of microtubules. Dynein is an ATPase attached to the microtubules and is required for ciliary movement. The apical cytoplasm of ciliated cells contains basal bodies (modified centrioles) for anchoring cilia and an abundance of mitochondria for producing ATP to drive ciliary beating. Mixtures of ciliated cells and goblet cells are found in much of the female reproductive tract including the uterine tubes and uterus and in the portions of the respiratory system proximal to the terminal bronchioles such as the trachea, bronchi and bronchioles. Terminal bronchioles have ciliated cells but lack goblet cells.

Examine the transmission electron micrograph below. Match the labeled structure in the micrograph with the most appropriate description of its microscopic anatomy or physiological role.

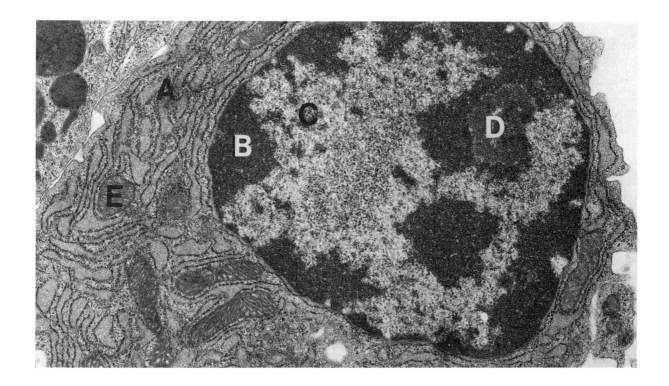

45. These contain nascent polypeptide chains destined for exocytosis.

46. These contain enzymes for the electron transport chain of oxidative phosphorylation.

47. This structure contains the DNA which encodes ribosomal RNA.

48. This cell is a

    (A) neutrophil
    (B) macrophage
    (C) plasma cell
    (D) lymphocyte
    (E) mast cell

49. The chief secretion product of this cell is

    (A) lipase
    (B) σ-amylase
    (C) immunoglobulin
    (D) transferrin
    (E) lactoperoxidase

Items 50-52

A 55 year-old alcoholic male shows hemorrhages near the base of hair follicles and swollen gingivae around his teeth. He has cutaneous lesions that have healed poorly. A fractured toe has not healed properly after 3 months. Histological examination of the lesions reveals extensive granulation tissue with few collagen fibers. A diagnosis of scurvy is made.

50. The cells most affected in scurvy are

    (A) liver parenchymal cells
    (B) osteoclasts
    (C) fibroblasts
    (D) plasma cells
    (E) eosinophils

51. Scurvy is the result of a defect in

    (A) elastin biosynthesis
    (B) tropocollagen biosynthesis
    (C) collagen cross-linking
    (D) laminin polymerization
    (E) fibronectin binding

52. The most accurate description of the fundamental lesion involved in the etiology of this dietary deficiency is

    (A) deficient synthesis of collagen alpha chains
    (B) leakage of capillaries
    (C) decreased cross-linkage at desmosyl residues
    (D) increased hydrolysis of tropocollagen telopeptides
    (E) increased collagen turnover

ANSWERS AND TUTORIAL ON ITEMS 45-49

The answers are: **45-A, 46-E, 47-D, 48-C, 49-C**. This is an electron micrograph of a plasma cell. Plasma cells differentiate from B-cells after appropriate antigenic stimulation and interaction with antigen presenting macrophages and helper T-cells. Plasma cells are highly differentiated cells dedicated to the synthesis and secretion of antibodies (immunoglobulins). Upon antigenic stimulation, the heterochromatinized nucleus of the B-lymphocyte decondenses to form a plasma cell nucleus with peripheral heterochromatin (B) and euchromatin (C). The nucleolus (D) contains the DNA sequences for the synthesis of ribosomal RNA, the essential component of ribosomes. Ribosomes bind to the endoplasmic reticulum to form the rough endoplasmic reticulum (A), the location where immunoglobulin mRNA is translated into the nascent polypeptide chains for immunoglobulin. Once synthesized, immunoglobulins are transported into the cisternae of the rough endoplasmic reticulum and then make their way out of the cell. Mitochondria (E) contain the enzymes for oxidative phosphorylation and electron transport crucial to the synthesis of ATP. Many of the anabolic functions of plasma cells are dependent on the energy-rich ATP produced in mitochondria.

ANSWERS AND TUTORIAL ON ITEMS 50-52

The answers are: **50-C, 51-C, 52-E**. Vitamin C deficiency results in a disease called scurvy. This disease is especially common in alcoholics because of dietary deficiency. Rupture of capillaries leads to follicular and gingival hemorrhages. Wound healing and fracture repair are also unusually slow in scorbutic patients. The granulation tissue at poorly healed wounds would have many fibroblasts, a cell type primarily involved in collagen biosynthesis. Vitamin C is an essential co-factor in enzymatic reactions catalyzing the hydroxylation of prolyl and lysyl residues of collagen. Procollagen molecules without hydroxyproline residues have significant instability in their triple helices and are therefore more susceptible to degradation. In addition, extracellular collagen molecules have few hydroxylysine residues. Their poor cross-linkage renders them more susceptible to turnover. Also, fibroblasts secrete collagen more slowly in the vitamin C deficient state. Bone growth and fracture repair are also abnormal in scurvy.

Items 53-56

Examine the scanning electron micrograph of a fractured specimen of liver tissue below. Match the labeled structure in the micrograph with the most appropriate microscopic anatomical or functional description of the labeled structure.

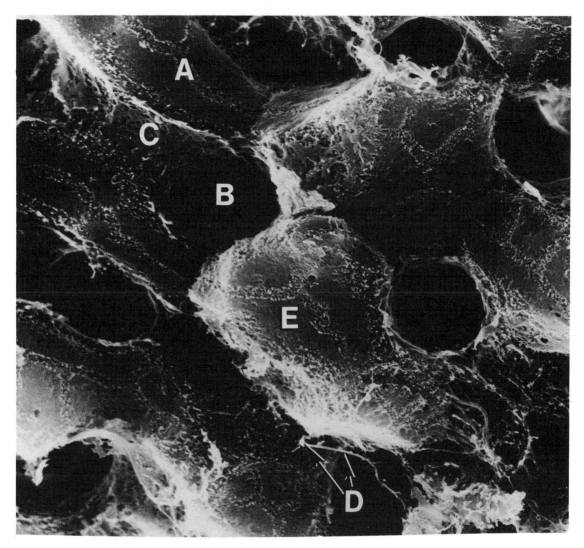

53. This structure receives blood from branches of the portal vein in the portal triads.

54. Liver parenchymal cells are joined by tight junctions at this structure to produce a blood-bile barrier.

55. This cell type is involved in synthesis of serum albumin and detoxification.

56. These endothelial cells form fenestrated and discontinuous capillaries.

ANSWERS AND TUTORIAL ON ITEMS 53-56

The correct answers are: **53-B, 54-A, 55-E, 56-C.** This is a scanning electron micrograph of fractured liver tissue. It consists numerous liver parenchymal cells (E) held together by delicate collagen fibers (D). Liver parenchymal cells are the main, functional, organ-specific cell type of the liver. They are involved in serum protein synthesis, metabolism of lipids and bile salts and detoxification of drugs. Individual liver parenchymal cells are joined together by tight junctions on surfaces facing the bile canaliculi (A). These minute channels convey bile toward the portal canal where they join with hepatic ducts. Cords of liver parenchymal cells are surrounded by sinusoids (B) which convey blood from the portal triads to the central vein of a liver lobule. The endothelial cells (C) lining these sinusoids are arranged in a fenestrated and discontinuous layer and have some phagocytic capacity.

Items 57-59

    (A) Rough endoplasmic reticulum
    (B) Smooth endoplasmic reticulum
    (C) Golgi apparatus
    (D) Nuclear envelope
    (E) Microvillus

For each organelle of the **placental syncytiotrophoblast**, select the associated cellular function.

57. Organelle that is involved in the final glycosylation of chorionic gonadotrophin.

58. Organelle that is involved in the synthesis of polypeptide chains for placental lactogen.

59. Organelle that is involved in progesterone synthesis.

ANSWERS AND TUTORIAL ON ITEMS 57-59

The answers are: **57-C, 58-A, 59-B.** The syncytiotrophoblast is the boundary between the maternal blood in the intervillous space of the placenta and the underlying fetal tissue of the placenta. The syncytiotrophoblast is formed by the proliferation and fusion of underlying cytotrophoblasts and is therefore a true syncytium. It has a complex ultrastructure that reflects the many different functions carried out in this syncytial epithelial layer. For example, the syncytiotrophoblast is the site of synthesis of human chorionic gonadotrophin (hCG), a glycoprotein hormone involved in maintaining the pregnant state of the female reproductive system during pregnancy. The polypeptides of hCG are synthesized in the rough endoplasmic reticulum (A) along with human placental lactogen and are glycosylated largely in the Golgi apparatus (C). Steroid biosynthesis occurs in part on the membranes of the smooth endoplasmic reticulum (B). The nuclear envelope (D) is a double unit membrane with octagonal pores. It serves as the boundary between the nucleus and the cytoplasm. Microvilli are prominent on the apical border of the syncytiotrophoblast. Here, like in all other apical brush borders, the microvilli mediate fluid, small molecule and large molecule transport across the placenta. Inorganic ions, sugars and amino acids are actively transported across the placenta through the microvilli. Maternal immunoglobulins are engulfed functionally intact by pinocytosis and transported in some of the membrane bound cytoplasmic vesicles (E) found everywhere in the syncytiotrophoblast.

Items 60-64

    (A) Simple squamous epithelium
    (B) Simple cuboidal epithelium
    (C) Simple columnar epithelium
    (D) Pseudostratified columnar epithelium
    (E) Stratified squamous epithelium

For each variety of epithelium, select the associated anatomical location where this type of epithelium is found in the human body.

60. Tracheal mucosa

61. Visceral pleura

62. Proximal convoluted tubule

63. Epidermis

64. Duodenal mucosa

ANSWERS AND TUTORIAL ON ITEMS 60-64

The answers are: **60-D, 61-A, 62-B, 63-E, 64-C.** Simple squamous epithelium (A) consists of a single layer of flattened epithelial cells. It is found at the lumen of all blood vessels (endothelium) and lining all serous cavities including the pleural cavities, pericardial cavity and peritoneal cavity (mesothelium). In all of these locations, the apical epithelial surface faces the lumen and prevents adhesion to maintain a patent lumen. In serous cavities, mesothelium secretes serous fluids and prevents adhesion of the visceral layer (on the organ) to the parietal layer (on the body wall). Simple cuboidal epithelium (B) consists of a single layer of cells that are equal in height and width. Simple cuboidal epithelium is found in proximal convoluted tubules, distal convoluted tubules and in the thyroid follicular epithelium. Simple columnar epithelium (C) consists of a single layer of cells that are taller than they are wide. This type of epithelium is present in the gastric and intestinal epithelium. Simple columnar epithelial cells often have a microvillous brush border that is involved in absorption of luminal contents. Pseudostratified columnar epithelium (D) consists of tall columnar epithelial cells that stretch all the way from the apex to the base of the epithelium and a variable population of cells resting on the basement membrane and reaching only part of the way up to the epithelial apex. It is found in the trachea, bronchi and large bronchioles of the respiratory system (where the tall columnar cells are ciliated) and is also widely distributed in the male reproductive system, e.g. in the epididymis, ductus deferens, prostate gland, seminal vesicles and parts of the urethra. Stratified squamous epithelium (E) is multilayered with flattened apical epithelial cells. It is found in the epidermis of the skin where it is keratinized and in the esophagus and vagina where it is unkeratinized. Stratified squamous epithelium is specialized to resist abrasion.

Items 65-69

    (A) Loose irregular connective tissue
    (B) Dense irregular connective tissue
    (C) Dense regular connective tissue
    (D) Hyaline cartilage
    (E) Elastic cartilage

For each variety of connective tissue, select the associated anatomical location where this type of connective tissue is found in the human body.

65. The articular surfaces of the humerus

66. The Achilles tendon

67. The auricle of the external ear

68. The reticular portion of the dermis

69. The lamina propria of the duodenum

24

The answers are: **65-D, 66-C, 67-E, 68-B, 69-A.** Loose irregular connective tissue (A) consists of many cells and a few randomly arranged fibers. It often contains resident cells (fibroblasts) and immigrant cells such as neutrophils, macrophages and plasma cells. It forms the mucosal connective tissue (lamina propria) of the duodenum and all other moist visceral organs. Dense irregular connective tissue (B) consists of a few cells and many densely packed fibers randomly arranged. It is most abundant in the deeper reticular layer of the dermis, the connective tissue of the skin. Dense regular connective tissue (C) consists of a few cells and many densely packed fibers regularly arranged. The tendons are classical examples of dense regular connective tissue. Cartilage is a kind of specialized connective tissue along with bone, bone marrow, blood, reticular tissue and adipose tissue. Hyaline cartilage (D) is abundant in the laryngeal and tracheal cartilages. It is also abundant in developing bones of the appendicular skeleton. Remnants of the cartilaginous models persist at the articular surfaces of long bones. Elastic cartilage (E) is quite similar to hyaline cartilage histologically but it contains many elastic fibers. It is found in the epiglottis, around the auditory tube, in laryngeal cartilages and in the auricle (pinna) of the external ear.

Items 70-74

(A) Neutrophil
(B) Eosinophil
(C) Basophil
(D) Monocyte
(E) Lymphocyte

For each variety of leucocyte, select the associated anatomical or functional description of this type of white blood cell found in whole human blood.

70. This cell differentiates into an immunoglobulin-secreting cell.

71. This cell is the most abundant granulocyte in blood.

72. This cell can differentiate into a microglial cell in the central nervous system.

73. This cell is abundant in patients with schistosomiasis.

74. This cell contains granules rich in heparin.

ANSWERS AND TUTORIAL ON ITEMS 70-74

The answers are: 70-E, 71-A, 72-D, 73-B, 74-C. The leucocytes of whole blood are either granulocytes (neutrophils, eosinophils and basophils in order of relative granulocyte frequency in peripheral blood) or agranulocytes (lymphocytes and macrophages, the former are the most abundant agranulocytes). All of the leucocytes are derived from bone marrow stem cells. Neutrophils (A) contain small granules that stain bluish in blood smears. They are involved in bacterial phagocytosis. Eosinophils (B) contain large bright red granules in blood smears. They are involved in combatting parasitic infestations such as found in schistosomiasis. Basophils (C) contain large bright purple granules in blood smears. Their granules contain heparin, an anticoagulant. Basophils serve to modulate the inflammatory response. Monocytes (D) are large agranulocytes with an irregular nucleus. They are migratory cells dedicated to phagocytosis and are part of the mononuclear phagocyte system located in many places throughout the body. Macrophages are found in peripheral blood (monocytes), in connective tissues (histiocytes), immune organs (where they are involved in phagocytosis and presentation of antigens), the lungs (alveolar macrophages), the liver (Kupffer cells) and the brain and spinal cord (microglia). Lymphocytes (E) are the primary cells of the immune system. They exist as either B-cells (plasma cell precursors-humoral immunity) or T-cells (modulators of the immune response-cellular immunity). Lymphocytes are abundant in the peripheral blood, lamina propria and immune organs such as lymph nodes, the thymus and the spleen.

Items 75-79

    (A) Mesenchymal fibroblast
    (B) Chondroblast
    (C) Osteoblast
    (D) Chondrocyte
    (E) Osteocyte

For each variety of connective tissue cell, select the associated anatomical or functional description of this type of specialized connective tissue cell found in the human body.

75. This deep periosteal cell can differentiate into a type I collagen-secreting cell.

76. This perichondrial cell can differentiate into a type II collagen-secreting cell.

77. This cell responds to parathormone by osteolysis.

78. This cell type aggregates to form centers of chondrification in embryonic limb buds.

79. This cell is capable of mitotic division when surrounded by type II collagen and cartilage-specific proteoglycan.

The answers are: **75-C, 76-B, 77-E, 78-A, 79-D**. Mesenchymal fibroblasts (A) are found in space-filling loose connective tissue in many embryonic locations including the developing limb buds. Here, they aggregate into centers of chondrification and then differentiate into chondroblasts and chondrocytes in cartilaginous models of bone. Chondroblasts (B) are found in the inner layers of the perichondrium and are capable of differentiating into chondrocytes (D), mature cells of cartilage that are surrounded by type II collagen and cartilage-specific proteoglycan. Osteoblasts (C) are found in the inner layers of the periosteum and all along the endosteum of bone. They secrete type I collagen and other matrix components that subsequently become calcified. Once osteoblasts have become entrapped in secreted calcified extracellular matrix, they are known as osteocytes. Osteocytes can both deposit calcium into new matrix under the influence of calcitonin or remove it from old matrix under the influence of parathormone.

Examine the transmission electron micrograph of skeletal muscle below. Match the labeled structure in the micrograph with the most appropriate description of its microscopic anatomy or physiological role.

80. These structures are the ends of individual sarcomeres. These are regions rich in *a*-actinin.

81. These structures are regions of overlap between actin-rich thin filaments and myosin-rich thick filaments.

82. These structures produce ATP for muscle contraction.

83. These structures are regions rich in actin-containing thin filaments where there is no overlap between thin and thick filaments.

84. These structures are regions rich in myosin-containing thick filaments where there is no overlap between thick and thin filaments.

ANSWERS AND TUTORIAL ON ITEMS 80-84

The answers are: **80-B, 81-C, 82-A, 83-D, 84-E.** This is a high power transmission electron micrograph of skeletal muscle. Skeletal muscle fibers contain many myofibrils, each consisting of many sarcomeres. Each sarcomere is formed by a regular array of thick (myosin-rich) and thin (actin-rich) filaments. Myofibrils are composed of many sarcomeres. Individual sarcomeres extend from Z line (B) to Z line. Sarcomeres have central A bands (C) where there is extensive overlap between thin, actin-rich filaments and thick, myosin-rich filaments. When a muscle fiber (cell) contracts, the width of the I bands (D) and H bands (E) decreases because the thin and thick filaments increase their overlapping due to sliding of thin filaments past thick filaments. The H bands are regions of no overlap between thin and thick filaments. H bands contain no thin filaments. The length of the A bands remains constant during muscle contraction. During muscle contraction, the I bands and H bands decrease in length and the Z lines move closer together, leading to a shortening of the sarcomere. When many sarcomeres shorten, the entire muscle cell shortens. Muscle contraction is driven by the energy rich compound ATP which is produced in mitochondria (A).

Contractile force in muscle is generated by a change in the position of actin and myosin, which is regulated by intracellular calcium concentration. Energy for muscle contraction is derived from the hydrolysis of ATP. Under appropriate conditions, the actin-myosin complex has ATPase activity. Release of energy from ATP hydrolysis causes conformational changes in the muscle proteins resulting in useful movement. A sarcomere has a variable total length depending on the contractile status of the cell. When a muscle contracts, thick and thin filaments slide past one another. During a contraction and relaxation cycle, calcium concentration around the myofibrils increases suddenly. This causes a conformational change in the troponin molecule, which exposes the S-1 (cross-bridge) binding site of actin, and a myosin-actin complex forms. Another conformational change occurs, and the S-1 fragment, still in association with the actin-containing thin filament, swings like an oar in an oarlock and causes the thin filament to slide relative to the thick filament. When this occurs at millions of cross-bridges, the entire sarcomere is shortened. The calcium concentration falls rapidly, following hydrolysis of ATP and the swinging of cross-bridges. The drop in calcium severs the association between actin and myosin and the contraction stops. ATP is hydrolysed to adenosine diphosphate (ADP), which subsequently is phosphorylated to form ATP. The hydrolysis and regeneration of ATP in muscle contraction explains the plethora of mitochondria in skeletal and cardiac muscle.

Items 85-87

Examine the transmission electron micrograph below and then choose the most appropriate answer.

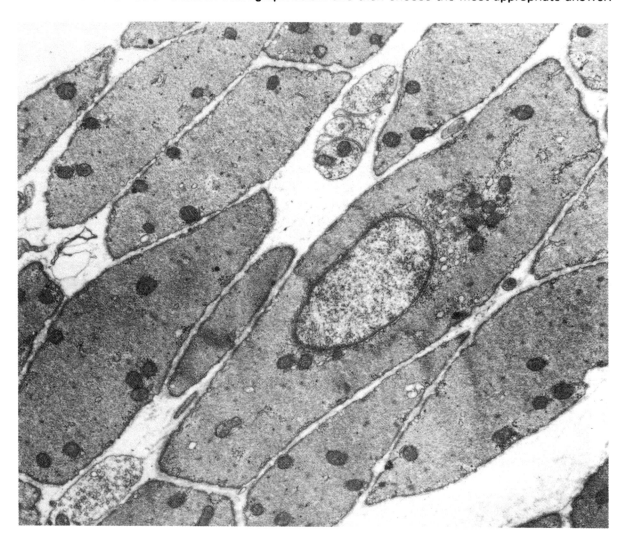

85. The predominant cell type in this electron micrograph is

   (A) fibroblast
   (B) chondrocyte
   (C) cardiac muscle cell
   (D) smooth muscle cell
   (E) adipocyte

86. This cell type is most abundant in

   (A) a tendon
   (B) hyaline cartilage
   (C) the cardiac ventricular walls
   (D) the wall of the urinary bladder
   (E) white fat

87. Which constituent would be most abundant in the cytoplasm of these cells?

    (A) Actin
    (B) Type I collagen
    (C) Type II collagen
    (D) Cholesterol esters
    (E) Cartilage proteoglycan

ANSWERS AND TUTORIAL ON ITEMS 85-87

The answers are: **85-D, 86-D, 87-A.** This is a low power electron micrograph of smooth muscle cells in the wall of the urinary bladder. Moist, hollow visceral organs including the urinary bladder have a luminal mucosa which consists of epithelium and lamina propria, a submucosa, a muscularis and an adventitial layer. When the adventitial layer is coated by a simple squamous epithelium (mesothelium), it is called the serosa. The muscularis in these visceral organs is rich in smooth muscle cells. These cells have a single nucleus per cell like cardiac muscle cells but lack the striations of cardiac muscle cells. Skeletal muscle cells have striations like cardiac muscle cells but have larger cells containing many nuclei (derived from many separate myoblasts) enclosed within a single plasma membrane (called the sarcolemma in the case of muscle cells). Smooth muscle cells have actin, myosin, troponin and tropomyosin but these contractile proteins are not arranged in well-organized sarcomeres. This is the reason that smooth muscle cells are not striated like skeletal muscle and cardiac muscle cells. The contractions of smooth muscle cells in the muscularis of hollow visceral organs regulates the diameter of the lumen. In the case of the urinary bladder, contraction of smooth muscle cells occurs when the full bladder contracts and empties urine into the urethra.

Examine the scanning electron micrograph of fractured tissue below and then answer the questions.

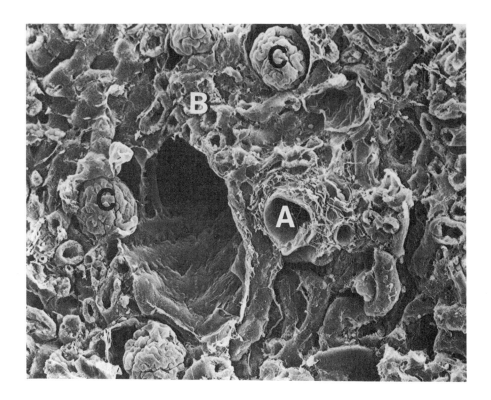

88. Structure A contains which of the following fluids?

    (A) Blood
    (B) Lymph
    (C) Bile
    (D) Dilute urine
    (E) Cerebrospinal fluid

89. Structure B has which of the following physiological roles?

    (A) Urine concentration
    (B) Nutrient and waste transport
    (C) Oxygen and CO_2 transport
    (D) Bile transport
    (E) Chylomicron transport

90. Structure C has which of the following physiological roles?

    (A) Nutrient absorption
    (B) Hormone synthesis
    (C) Blood filtration to remove nitrogenous wastes
    (D) Gas transport
    (E) Lymphocyte transport

91. This morphological arrangement is found in the

    (A) adrenal cortex
    (B) liver parenchyma
    (C) brain parenchyma
    (D) thymic cortex
    (E) renal cortex

ANSWERS AND TUTORIAL 88-91

The answers are: **88-A, 89-A, 90-C, 91-E.** This is a low power scanning electron micrograph of the renal cortex. The cortical portion of the kidneys consists of large numbers of blood vessels (A), renal corpuscles (C) and tubules (B). Large branches of the renal arteries carry blood rich in nitrogenous wastes into the afferent arterioles which supply a glomerular capillary tuft in the renal corpuscles. The glomerular capillary tuft resides in a deep invagination of Bowman's capsule, the other portion of the renal corpuscle. Bowman's capsule has a visceral epithelial layer consisting of podocytes and a parietal layer consisting of simple squamous epithelial cells that are continuous with the cuboidal epithelial cells of the proximal convoluted tubule. The glomerular blood filtrate crosses the glomerular basement membrane, passes between the foot processes of podocytes and enters the urinary space of the renal corpuscle. The urinary pole of Bowman's capsule conveys dilute urine into the proximal convoluted tubules where the complex process of recovery of proteins and salts from the blood filtrate begins. As the filtrate passes through the proximal convoluted tubules, loop of Henle, distal convoluted tubules and collecting tubules, the blood filtrate is modified further by recovery of most of the protein that crosses the glomerular basement membrane and by changes in salt and urea concentrations to produce urine.

Examine the scanning electron micrograph below and then choose the correct answer.

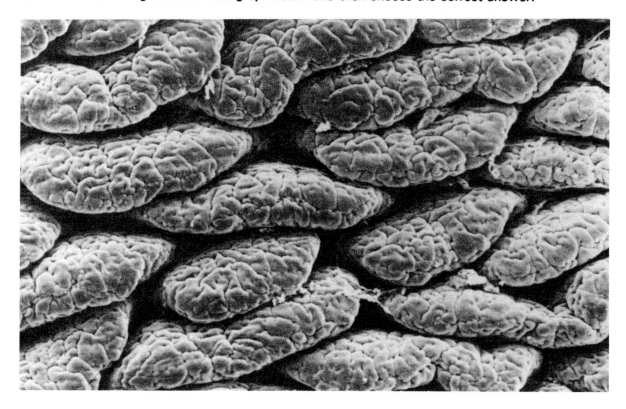

92. These structures have which major physiological function?

    (A) Taste perception
    (B) Nutrient absorption
    (C) Bile concentration
    (D) Gamete transport
    (E) Blood transport

93. These structures are characteristic of the mucosal surface of which organ?

    (A) Stomach
    (B) Jejunum
    (C) Uterine tube
    (D) Tongue
    (E) Gall bladder

94. The mucosal epithelium here consists of

    (A) ciliated and secretory cells
    (B) ciliated cells only
    (C) absorptive cells and goblet cells
    (D) a pseudostratified ciliated columnar epithelium with goblet cells
    (E) transitional epithelium

Examine the high power transmission electron micrograph of the junctions between three liver parenchymal cells and then choose the correct answer.

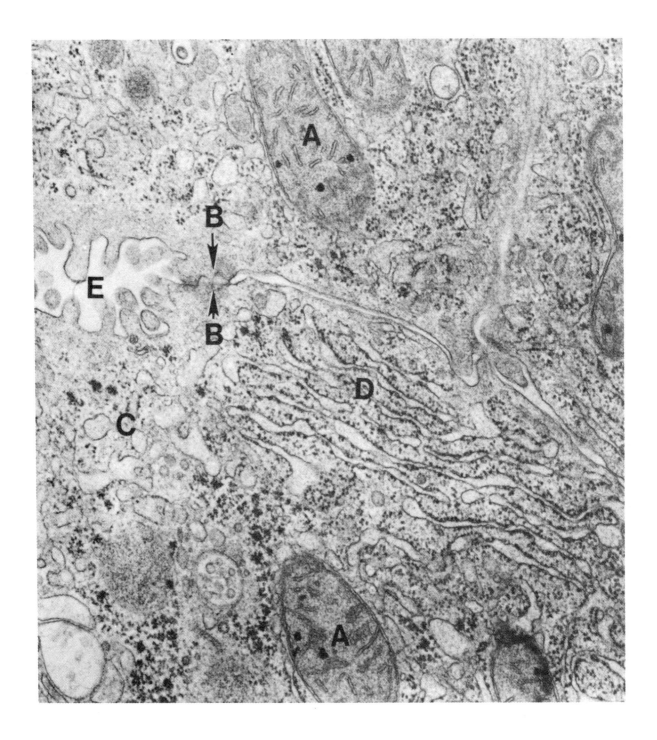

95. This structure contains enzymes for the electron transport chain and oxidative phosphorylation.

96. This structure forms a tight anatomical junction which serves as the blood-bile barrier.

97. This structure is involved in the initial stages of the synthesis of serum albumin.

98. This structure conveys bile toward the hepatic ducts.

ANSWERS AND TUTORIAL ON ITEMS 92-94

The answers are: **92-B, 93-B, 94-C.** This is a scanning electron micrograph of the mucosal surface of the jejunum, the second longest component of the small intestine. While some digestion and nutrient absorption occurs in the stomach, most of the digestion and nutrient absorption occurs in the small intestine. The mucosa of the jejunum has a plethora of villi. Jejunal villi are long flat structures covered by a simple columnar epithelium consisting of many tall, thin, columnar absorptive cells with an apical microvillous brush border. These cells are primarily responsible for absorption of digested nutrients from the lumen of the small intestine. In addition, the jejunal mucosal epithelium has goblet cells interspersed between columnar absorptive cells. Goblet cells secrete a thick coat of mucus that protects the jejunal mucosa from digestion.

ANSWERS AND TUTORIAL ON ITEMS 95-98

The answers are: **95-A, 96-B, 97-D, 98-E.** This is a high power transmission electron micrograph of a part of three liver parenchymal cells. Mitochondria (A) have enzymes for the electron transport chain and oxidative phosphorylation. These enzymes are used to synthesize ATP. This energy rich compound is utilized for many of the anabolic processes occurring in the liver. For example, the liver is actively engaged in protein synthesis. The rough endoplasmic reticulum (D) is the site where mRNAs are translated into polypeptide chains. Bile constituents are conjugated in the smooth endoplasmic reticulum (C) and are secreted into the bile canaliculi (E) between parenchymal cells. Tight junctions (B) surround the bile canaliculi and prevent bile from leaking into the vascular spaces in the liver. Thus, these tight junctions represent the anatomical basis for the blood-bile barrier.

Items 99-102

Examine this high power transmission electron micrograph below. It is an array of long hollow structures cut perpendicular to their long axis. The diameter of each hollow structure is 25 nm.

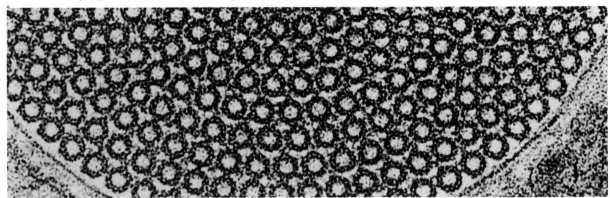

99. Each individual circular structure is a

    (A) microfilament
    (B) ciliary axoneme
    (C) microtubule
    (D) thick filament of muscle
    (E) intermediate filament

100. The most abundant protein found in these structures is

    (A) tubulin
    (B) actin
    (C) keratin
    (D) desmin
    (E) myosin

101. Which of the following proteins is present in abundance?

    (A) Actin
    (B) Vimentin
    (C) Microtubule associated proteins
    (D) α-actinin
    (E) Tropomyosin

102. Which description is most appropriate for these structures?

    (A) The thick filaments are composed of aggregates of myosin molecules.
    (B) Dynein rich side-arms are involved in their movement.
    (C) They consist of globular actin molecules arranged in two intertwined helices with troponin and tropomyosin molecules arrayed along the helix.
    (D) They consist of 13 protofilaments composed of alternating α- and ß-tubulin subunits arranged like a string of beads.
    (E) They are abundant in the cores of microvilli.

Examine the high power transmission electron micrograph of cell surface projections below. These projections are cut perpendicular to their long axis. Match the most appropriate description of the functional role of the structure with the labeled structure in the electron micrograph.

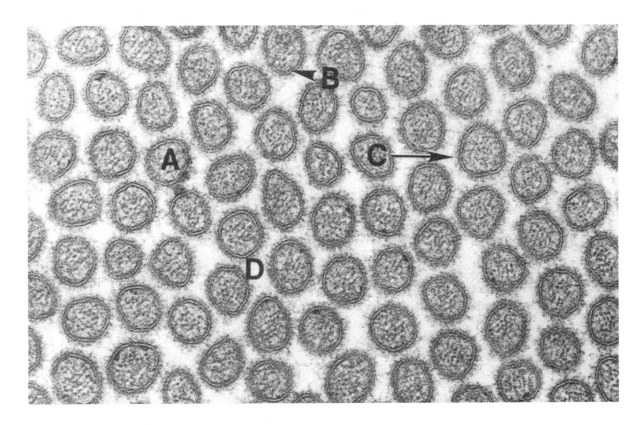

103. These structures are actin-rich microfilaments. They have a structural role.

104. These structures are a phospholipoprotein bilayer serving as a selective permeability barrier surrounding the entire cell.

105. This is a glycoprotein rich layer containing enzymatic activities for the hydrolysis of disaccharides.

106. These surface projections have which primary physiological function?

    (A) Movement of luminal contents
    (B) Transduction of mechanical into electrical energy
    (C) Transduction of chemical into electrical energy
    (D) Absorption of luminal contents
    (E) Gas exchange

ANSWERS AND TUTORIAL ON ITEMS 99-102

The answers are: **99-C, 100-A, 101-C, 102-D.** This is a high power transmission electron micrograph of an array of microtubules cut perpendicular to their long axis. Microtubules are long hollow structures with an outside diameter of 25 nm. They consist of 13 protofilaments. Each protofilament consists of alternating subunits of α- and ß-tubulin. Tubulins have similar molecular weights (55 kD) but are distinct due to subtle chemical differences. In addition to tubulin, microtubules have microtubule associated proteins that are involved in the interaction of microtubules with other cytoskeletal elements. Microtubules are important cytoskeletal structures that are involved in maintenance of cell morphology. They are also important constituents of cilia, flagella, centrioles and the mitotic spindle. Microtubules are important for moving entire cells as well as chromosomes within cells during mitosis.

ANSWERS AND TUTORIAL ON ITEMS 103-106

The answers are: **103-A, 104-B, 105-C, 106-D.** This is a high power transmission electron micrograph of sections cut perpendicular to the long axis of the microvilli in the jejunum of the small intestine. Microvilli are apical surface projections designed to increase the cell surface area. They are particularly well developed in areas where luminal contents are being absorbed, e.g. in the small and large intestine and proximal and distal convoluted tubules in the kidneys. Like all other cell surface projections, microvilli are covered by a phospholipoprotein bilayer of plasma membrane (B). The cores of microvilli are filled with many microfilaments (A) with a diameter of 6 nm. These microfilaments are rich in actin and are involved in movement of microvilli. The plasma membrane of intestinal microvilli has a thick glycocalyx (C) which consists of the glycoprotein rich extracellular domains of integral membrane proteins. This fuzzy coat on the outer leaflet of the plasma membrane consists of minute filaments, 2,5-5 nm in diameter. They often project 0.1 to 0.5 μm beyond the apical tips of microvilli. The glycocalyx prevents digestive enzymes in the lumen of the small intestine from gaining access to epithelial cells. In addition, the glycocalyx contains digestive enzymes that complete the final steps in nutrient digestion.

Examine the light micrograph below and then choose the most appropriate labeled structure to match the functional role or morphological description of this structure.

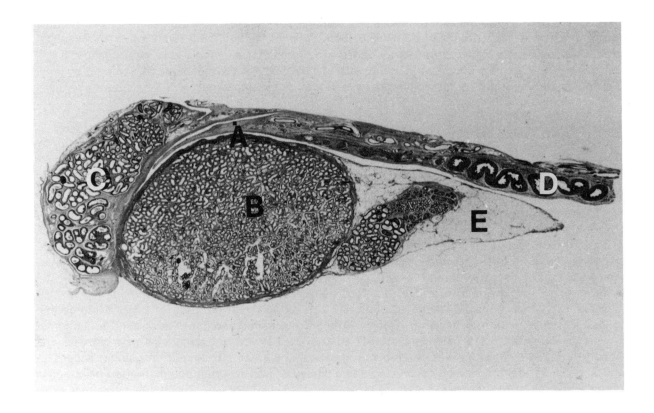

107. This structure has a pseudostratified epithelial lining and a thick coat of smooth muscle. It leads directly into the ejaculatory ducts.

108. This thick connective tissue capsule has septa that divide its enclosed tissue into lobules.

109. This structure is lined by a pseudostratified epithelium with long stereocilia. Spermatozoa undergo maturation here.

110. This structure contains both Sertoli cells and Leydig cells. Spermatogenesis occurs here.

ANSWERS AND TUTORIAL ON ITEMS 107-110

The answers are: **107-D, 108-A, 109-C, 110-B.** This is a low power light micrograph of a sagittal section through the testis (B), epididymis (C) and ductus deferens (D). The testis is encapsulated by a dense fibrous connective tissue capsule called the tunica albuginea (A). Septa penetrate deep into the testis from the tunica albuginea and divide each testis into several hundred testicular lobules. Each lobule consists of many seminiferous tubules. Spermatogenesis, a process where proliferative spermatogonia differentiate into primary spermatocytes and eventually haploid spermatozoa, occurs in the seminiferous tubules of the testis. Meiosis begins in the primary spermatocyte, eventually resulting in the formation of millions of haploid spermatozoa. The seminiferous tubules are surrounded by interstitial tissue rich in connective tissue fibroblasts, fenestrated capillaries and testosterone-secreting Leydig cells. Spermatogenesis and the secretory activities of the excurrent duct system of the testes are testosterone-dependent. The seminiferous tubules empty into the rete testis which in turn empties into the efferent ductules. The efferent ductules join the head of the epididymis, a long convoluted tube interconnecting the efferent ductules and the ductus deferens. The mucosal epithelium of the epididymis and ductus deferens are both pseudostratified columnar. Spermatozoa are thought to mature in the epididymis. They are conveyed through the ductus deferens into the prostate and out the penile urethra during ejaculation.

<u>Items 111-114</u>

    (A) Macrophage
    (B) Helper T-cell
    (C) Natural killer cell
    (D) B-cell
    (E) Plasma cell

For each cell of the immune system, select the most appropriate functional role in the immune response.

111. In conjunction with antigen-presenting and antibody producing cell, this cell is required for antibody synthesis.

112. This cell is a bone marrow derivative involved in presentation of antigen to immunoglobulin producing cell.

113. This cell has an abundant rough endoplasmic reticulum and a prominent nucleolus. It dedicated to immunoglobulin synthesis and secretion into the blood.

114. This cell has a granular cytoplasm with azurophilic granules. It is a differentiated cell which represents the first line of defense against foreign cells.

ANSWERS AND TUTORIAL ON ITEMS 111-114

The answers are: **111-B, 112-A, 113-E, 114-C.** The immune system contains a diverse assortment of cells for defense against foreign antigens and foreign cells. In the humoral immune response, B-cells (D) differentiate into immunoglobulin secreting plasma cells (E). Macrophages (A) are derived from bone marrow and in conjunction with helper T-cells (B) must present foreign antigens to B-cells to control their differentiation into plasma cells. Once B-cells are stimulated to differentiate into plasma cells, their nucleus becomes less condensed, a nucleolus appears for synthesis of ribosomal RNA and the cytoplasmic compartment of the cell expands to accommodate an increase in rough endoplasmic reticulum and Golgi apparatus. The cytoplasmic machinery for immunoglobulin synthesis and secretion thus appears as a B-cell differentiates into a plasma cell. Our immune system also contains a population of natural killer (NK) cells (C). NK cells are highly differentiated cell types that serve as the first line of defense against foreign cells. They are cytolytic entities and can act directly on foreign cell types, thus destroying them rapidly. The formation of killer T-cells from small lymphocytes in response to foreign cells requires a delay, during which an abnormal tumor cell might proliferate more rapidly than the killer T-cells designed to combat the transformed cell. When a tumor arises, NK cells probably attack it first and limit the further development of tumor cells until a new cadre of killer T-cells forms a second line of defense against the malignant tumor cells. They are larger than lymphocytes (12-15 μm), have a lobulated nucleus and many cytoplasmic azurophilic granules.

<u>Items 115-118</u>

  (A) Lymph node
  (B) Spleen
  (C) Liver
  (D) Thymus
  (E) Bone Marrow

For each organ of the immune system, select the most appropriate functional role in the immune response.

115. This organ is derived from pharyngeal pouches III and IV. It contains reticular epithelial cells and lymphocytes.

116. This organ has a subcapsular sinus. Germinal centers appear in response to antigenic stimulation.

117. This organ is responsible for immune surveillance of the blood and removal of effete erythrocytes from the systemic circulation.

118. This organ is the first major hematopoietic organ in the embryo.

The labeled light micrographs below are taken from the deep (left) and superficial (right) portions of the fundic gastric pits. Choose the most appropriate description of the morphology or functional role of the labeled cell type.

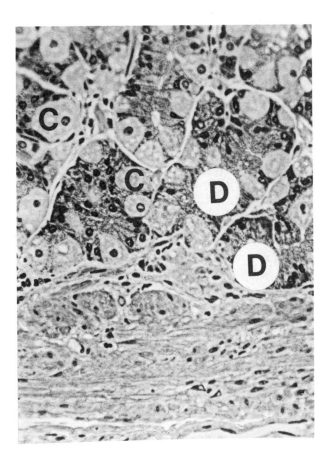 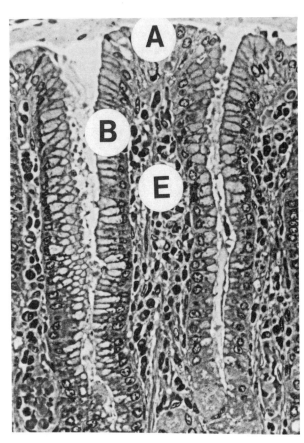

119. These cells are acidophilic, have an extensive apical canalicular system and have many mitochondria closely associated with the canaliculi.

120. These cells have a basophilic cytoplasm rich in rough endoplasmic reticulum. They secrete pepsinogen.

121. These mucous surface cells secrete a protective layer of mucus and form the most superficial cells of the gastric glands.

122. These cells secrete gastric intrinsic factor.

123. The secretion of these cells is sensitive to gastrin.

ANSWERS AND TUTORIAL ON ITEMS 115-118

The answers are: **115-D, 116-A, 117-B, 118-C.** Lymph nodes (A) are distributed widely throughout the body. They are especially abundant in the cervical, axillary and inguinal regions where they receive the lymphatic drainage of the head, upper extremities and lower extremities respectively. Lymph nodes receive lymph via peripheral afferent lymphatics. These vessels convey lymph into the subcapsular sinuses and then allow it to circulate slowly over germinal centers distributed throughout the cortex. Germinal centers contain macrophages, lymphocytes and plasma cells. After percolating past germinal centers, lymph drains into sinuses in the medullary portion of the node and exits via the efferent lymphatics in the hilus. The lymph nodes are responsible for the immune surveillance of the lymph. The spleen (B) is located in the dorsal mesentery of the stomach. It consists of white pulp (periarterial lymphatic sheaths = PALS) and red pulp. The PALS contain germinal centers and are responsible for production of antibodies directed against blood-borne antigens. The red pulp consists of a complex network of sinusoids that entrap and destroy aged and damaged erythrocytes. The liver (C) is a relatively large organ in the embryo. It serves as the major hematopoietic organ after the yolk sac regresses and before the bone marrow (E) develops into the predominant site of hematopoiesis. The thymus (D) is a lobulated organ consisting of reticular epithelial cells derived from the IIIrd and IVth pharyngeal pouches (endoderm) and large numbers of lymphocytes. The reticular epithelial cells secrete a protein called thymosin which is required for differentiation of T-cells. T-cells formed in the thymus then leave the organ and are dispersed widely throughout the blood, peripheral connective tissues, lymph nodes and spleen.

ANSWERS AND TUTORIAL ON ITEMS 119-123

The answers are: **119-C, 120-D, 121-A, 122-C, 123-C.** This is a light micrograph of the gastric mucosa which has many glandular pits throughout the fundic and corpic regions. These glands have several different kinds of epithelial cells. The most superficial portion of gastric glands contains mucous surface cells (A). These cells extend a short way into the upper portion of gastric glands but are soon replaced by mucous neck cells (B). Both cells are similar in their structure but secrete mucus with differing chemical properties. Their secretions coat the gastric mucosa and prevent its ulceration by the proteolytic enzymes inside the gastric lumen. Beneath the mucous neck cells, one encounters parietal cells (C) and chief cells (D). Parietal cells are more abundant in the upper portions of the gastric glands. They have a complex apical canalicular system that serves to increase their surface area. These cells also have a large number of acidophilic mitochondria closely associated with the canalicular system. Parietal cells secrete HCl and gastric intrinsic factor. Enteroendocrine cells known as G cells secrete gastrin, a polypeptide that stimulates HCl secretion from parietal cells. Gastric intrinsic factor is required for absorption of vitamin B_{12}. Chief cells are more abundant in the deepest portions of gastric glands. Chief cells have a fine structure that is appropriate for the synthesis and secretion of the main digestive enzyme of the stomach, pepsinogen. Thus, they have a basophilic cytoplasm with an abundance of rough endoplasmic reticulum, abundant mitochondria, a well developed Golgi apparatus and zymogen granules containing pepsinogen. Once secreted, pepsinogen is cleaved to pepsin, a form with proteolytic activity and a pH optimum around 2.0. The gastric mucosa also has a lamina propria consisting of a loose areolar connective tissue rich in fibroblasts (E).

Examine the labeled light micrograph below and then choose the most appropriate description of the morphology or functional role of the labeled structure.

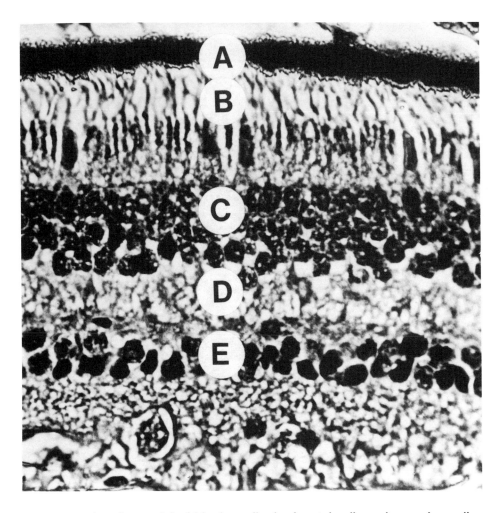

124. This structure contains the nuclei of bipolar cells, horizontal cells and amacrine cells.

125. This structure contains the nuclei of rods and cones.

126. This structure is rich in melanin. The apical surfaces of epithelial cells here phagocytose the apical portion of effete rod and cone outer segments.

127. This structure contains neuronal processes connecting rods and cones to bipolar cell.

Items 128-131

Examine the labeled light micrograph below and then choose the most appropriate description of the morphology or functional role of the labeled structure.

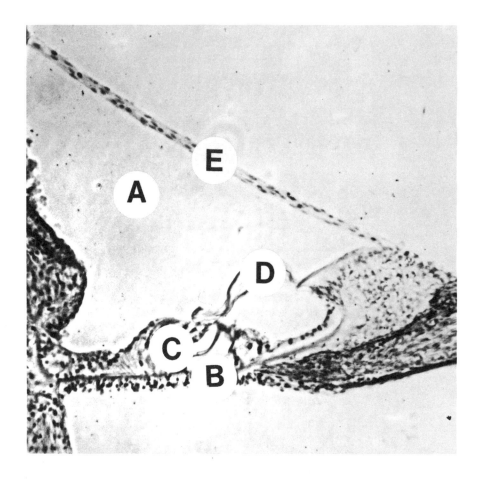

128. The apical surfaces of hair cells contact this structure.

129. This structure contains a blood filtrate which is produced in the stria vascularis and resorbed in the endolymphatic sac.

130. The apical projections from this structure are highly modified microvilli.

131. Basilar fibers are located in this structure.

ANSWERS AND TUTORIAL ON ITEMS 124-127

The answers are: **124-E, 125-C, 126-A, 127-D.** This is a photomicrograph of the photoreceptive portion of the neural retina. The innermost layer here consists of the pigmented retina (A). This is a simple cuboidal epithelium of melanin-containing pigmented cells. It functions to absorb stray light and also phagocytoses the apical portions of the outer segments of the rods and cones (B). The nuclei of the rods and cones are contained in the outer nuclear layer (C). Axons of these rods and cones project through the outer plexiform layer (D) and synapse with dendrites of bipolar cells, the nuclei of which are found in the inner nuclear layer (E) along with the nuclei of Müller cells, horizontal cells and amacrine cells. Müller cells are supportive, tall columnar cells that span much of the thickness of the neural retina. Horizontal cells integrate signals from multiple photoreceptors. Amacrine cells establish contacts between ganglion cells. Their function is poorly understood. Processes extending between the inner nuclear layer and the ganglion cell layer are found in the inner plexiform layer. Ganglion cells have their nuclei in the ganglion cell layer and project axons toward the optic nerve through the optic nerve fiber layer.

ANSWERS AND TUTORIAL ON ITEMS 128-131

The answers are: **128-D, 129-A, 130-C, 131-B.** This is a light micrograph of the organ of Corti, the mechanoreceptor in the inner ear responsible for converting vibrations into electrical impulses for sound perception. The organ of Corti resides within an endolymph-filled cavity called the cochlear duct (scala media) (A). Endolymph is a blood filtrate secreted into the cochlear duct at the stria vascularis and resorbed in a diverticulum of the membranous labyrinth called the endolymphatic sac. The organ of Corti rests on the basilar membrane (B). Basilar fibers reside in the basilar membrane. Traveling waves of deflection are produced in the basilar membrane by vibrations in the oval window. These waves are damped out at different locations along the basilar membrane depending upon the frequency of the traveling waves, resulting in stimulation of different hair cells (C) depending upon the frequency of the sound. Hair cells have long apical microvilli that contact the relatively stationary tectorial membrane (D). When the basilar membrane moves the hair cells, their apical microvilli are deformed because they move with respect to the relatively fixed tectorial membrane. These deformations lead to the initiation of action potentials from hair cells. Subsequently, these action potentials are conducted into the brain where they are perceived as sounds. The vestibular membrane (E) is the boundary between the cochlear duct and the scala vestibuli. The basilar membrane is the boundary between the cochlear duct and the scala tympani. The scala vestibuli and the scala tympani are continuous with one another. Both are filled with perilymph.

    (A) Thyroid Gland
    (B) Parathyroid Gland
    (C) Islets of Langerhans
    (D) Adrenal Medulla
    (E) Adrenal Cortex

Choose the best anatomical description of the endocrine gland.

132. This gland has parenchymal epithelial cells derived from pharyngeal pouch endoderm and secretes a hormone that stimulates osteocytic osteolysis.

133. This gland has neural crest cell derivatives that secrete hormones that increase pulse and respiratory rate.

134. This gland has parenchymal epithelial cells derived from coelomic epithelium. These epithelial cells have numerous cytoplasmic lipid droplets and an abundance of smooth endoplasmic reticulum.

Items 135-137

    (A) Esophagus
    (B) Fundic Stomach
    (C) Duodenum
    (D) Ileum
    (E) Colon

Choose the best microscopic anatomical description of the digestive system organ.

135. Mucosa has simple columnar epithelium of absorptive and goblet cells. Submucosa has numerous mucous glands. Muscularis externa consists of two complete bands of smooth muscle.

136. Mucosa has stratified squamous unkeratinized epithelium. Muscularis externa contains a mixture of smooth and skeletal muscle.

137. Mucosa contains deep pits with mucus and pepsinogen secreting cells. Submucosa lacks glands.

ANSWERS AND TUTORIAL ON ITEMS 132-134

The answers are: **132-B, 133-D, 134-E.** The thyroid gland (A) consists of follicular epithelial cells that secrete thyroglobulin and parafollicular (C-cells) that secrete calcitonin. Thyroid follicular epithelial cells are derived from a foregut thyroid diverticulum and are therefore of endodermal origin. The C-cells are neural crest derivatives. They secrete an antagonist of parathormone, the chief secretion product of the chief cells of the parathyroid glands (B).

The parathyroid glands are derived from the IIIrd and IVth pharyngeal pouches and are therefore endodermal in origin. Parathormone secretion results in elevation of serum calcium levels due to increased bone resorption, primarily by osteolytic activity of osteocytes.

The islets of Langerhans are endocrine tissue within the pancreas. They are endodermal derivatives like the pancreatic acinar cells. The islet tissue secretes insulin and glucagon. The adrenal gland consists of a medullary portion derived from neural crest and a cortical portion derived from coelomic epithelium (mesoderm). The adrenal medulla (D) contains two major populations of cells. One secretes epinephrine and the other secretes norepinephrine. These catecholamines are involved in the flight or fight response and result in increase in blood pressure, pulse and respiratory rate. The adrenal cortex (E) consists of several layers of cells dedicated to steroid hormone synthesis. Many cells of the adrenal cortex have large numbers of cytoplasmic lipid droplets (where steroid precursor cholesterol esters are stored) and an abundance of cytoplasmic smooth endoplasmic reticulum. Many of the enzymes for steroid synthesis are localized in the membranes of the smooth endoplasmic reticulum.

ANSWERS AND TUTORIAL ON ITEMS 135-137

The answers are: **135-C, 136-A, 137-B.** The esophagus (A) is a tube which connects the pharynx and the stomach. It conveys food to the stomach. It has a stratified squamous unkeratinized mucosal epithelium with deep mucous glands in the mucosa and submucosa. The muscularis externa of the esophagus is a mixture of skeletal and smooth muscle in the upper portion. The fundic stomach consists of deep gastric glands with mucus-secreting cells and chief cells that secrete pepsinogen, a proteolytic enzyme with an acidic pH optimum. Gastric glands also have parietal cells responsible for secreting HCl and gastric intrinsic factor (which is essential for absorption of dietary vitamin B_{12}). The muscularis externa of the stomach has several layers of smooth muscle obliquely arranged. The duodenum (C) of the small intestine has a mucosal epithelium consisting of a mixture of tall columnar absorptive cells and mucus-secreting goblet cells. Submucosal Brunner's glands are restricted to the duodenum. They secrete bicarbonate and mucus to neutralize acidic gastric chyme and prevent ulceration of the duodenal mucosa. The muscularis externa of the duodenum and ileum (D) consists of two robust layers of smooth muscle. The inner layer is arranged in a circular fashion and the outer layer is longitudinal. The colon (E) has a mucosa with deep pits lined by a mixture of columnar absorptive and goblet cells. The muscularis externa of the colon has a continuous inner circular band of smooth muscle and three robust strips of longitudinal smooth muscle called the taenia coli.

CHAPTER II

EMBRYOLOGY

Items 138-143

A 15 year old adolescent presents with a chief complaint of amenorrhea. She is 5'9" tall and weighs 130 lbs. Her temperature is 98.7° F., pulse is 64 beats/min. and blood pressure is 120/60. Physical examination reveals a normal female body habitus except for scant axillary and pubic hair. Pelvic examination reveals a shallow vagina with no cervix. A mobile mass can be palpated in the left labium majorum. Ultrasonographic examination reveals a complete lack of uterus and adnexal structures.

138. The most appropriate subsequent test to perform for your differential diagnosis is

   (A) laparoscopy
   (B) MRI of pelvis
   (C) karyotype analysis
   (D) glucose tolerance test
   (E) cardiac stress test

139. Which of the following chromosomal complements would you be most likely to find with this patient after karyotype analysis?

   (A) 45, X
   (B) 47, XX, +21
   (C) 47, XXY
   (D) 46, XY
   (E) 69, XXY

140. The most appropriate diagnosis is

   (A) Turner Syndrome
   (B) Down Syndrome
   (C) Klinefelter Syndrome
   (D) testicular feminization syndrome
   (E) triploidy

141. The most appropriate follow-up procedure indicated is

    (A) treatment with androgens
    (B) treatment with estrogens
    (C) ovariectomy
    (D) tubal ligation
    (E) orchiectomy

142. This patient would lack paramesonephric duct derivatives because

    (A) testes produce müllerian inhibiting substance
    (B) their differentiation is testosterone dependent
    (C) the patient is genetically female
    (D) the patient is genetically male
    (E) no testis determining factor is produced

143. All of the following statements concerning this patient are true EXCEPT:

    (A) H-Y antigen present
    (B) sterility
    (C) male gender identity
    (D) testis determining factor present
    (E) increased risk of seminoma

ANSWERS AND TUTORIAL ON ITEMS 138-143

The answers are: **138-C; 139-D; 140-D; 141-E; 142-A; 143-C.** The most appropriate next test would be a karyotype analysis followed by measurements of serum steroids. You would discover that the patient had a 46, XY normal male karyotype and normal testosterone levels. Further tests would reveal a deficiency in testosterone receptor. This patient has testicular feminization syndrome. There is an increased risk of seminoma in undescended testes and orchiectomy would be appropriate for the left testicle in the left labium majorum and the right testicle in the inguinal canal or body cavity. While genetically male, this patient would have a female gender identity but would be sterile because of a lack of ovaries and hyalinization of seminiferous epithelium in undescended testes. Because of the presence of a Y chromosome, H-Y antigen and testis determining factor would be present. Paramesonephric duct derivatives (uterine tubes, uterus and upper portion of vagina) would be absent due to müllerian inhibiting substance produced by the testes.

Items 144-148

A 28 year-old woman of Irish ethnic background in the 30th week of her first pregnancy reports a lack of fetal movements. Maternal serum α-fetoprotein levels are significantly elevated. Ultrasonography reveals the following (FC = fetal cranium; FCS = fetal cervical spine):

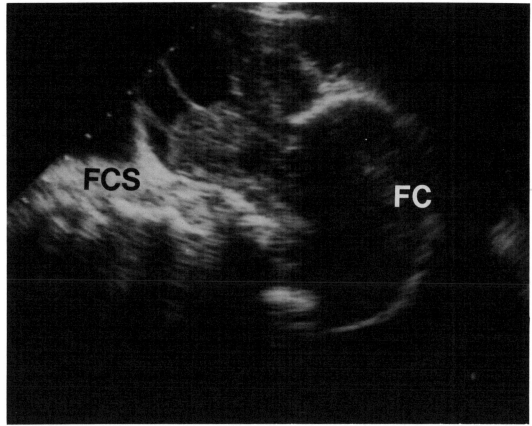

144. The most correct diagnosis of this anomaly is

   (A) anencephaly
   (B) spina bifida
   (C) respiratory distress syndrome
   (D) spina bifida occulta
   (E) encephalocele

145. All of the following statements are true concerning this anomaly EXCEPT:

   (A) caused by point mutation
   (B) likelihood of recurrence of similar congenital anomaly in subsequent pregnancies elevated
   (C) if carried to term, child has significant risk of paralysis
   (D) it is a variety of neural tube defect
   (E) fetus will live after birth

146. The most appropriate subsequent test for confirmation of diagnosis is

   (A) assay for amniotic fluid lecithin/sphingomyelin ratio
   (B) assay for amniotic fluid prostaglandin
   (C) assay for amniotic fluid α-fetoprotein
   (D) karyotype analysis of fetal cells
   (E) karyotype analysis of maternal cells

147. The most likely developmental process causing this congenital anomaly is

   (A) failure of neural tube closure
   (B) failure of lung differentiation
   (C) duodenal atresia
   (D) breakage of chromosomes
   (E) non-disjunction during meiosis

148. The abnormal process causing failure of formation of the skull most likely is

   (A) abnormal bone induction
   (B) excessive production of cerebrospinal fluid
   (C) inadequate fetal growth
   (D) formation of amniotic bands
   (E) abnormal placental transport of calcium

ANSWERS AND TUTORIAL ON ITEMS 144-148

The answers are: **144-E, 145-A, 146-C, 147-A, 148-A.** This ultrasonogram shows an example of an encephalocele. This is a type of neural tube defect. Neural tube defects show a complex multifactorial etiology with both genetic and environmental factors contributing to an increased incidence. In this case, the base of the occipital bone is poorly formed. During early development, the neural tube induces formation of the bony structures protecting the central nervous system such as the neural arches and the vault of the skull. In many types of neural tube defects, there is both abnormal formation of nervous tissue leading to neurological deficits and abnormal formation of the overlying bony structures normally protecting the nervous system. In encephalocele, meninges protrude into the space created by abnormal formation of the base of the occipital bone. Fetal blood and cerebrospinal fluid contains α-fetoprotein. In many neural tube defects, this α-fetoprotein leaks into the amniotic fluid (where it can be measured as an indication of neural tube defects) and may cross into the maternal circulation and subsequently be detected by an assay of maternal serum α-fetoprotein.

Examine the karyotype below and then answer the questions.

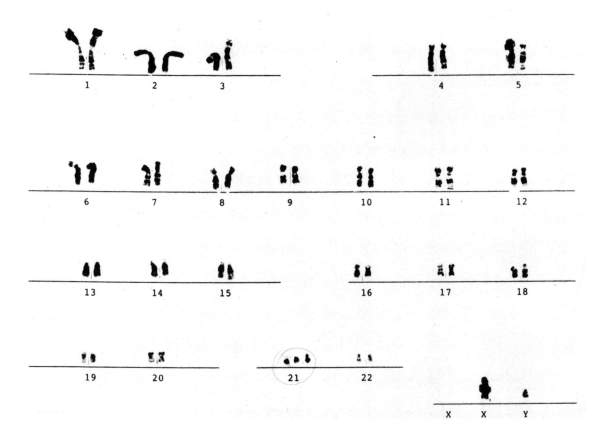

149. The developmental process whose disruption produces this karyotype is

    (A) mitosis
    (B) synapsis
    (C) differentiation of primordial germ cells
    (D) meiotic disjunction
    (E) migration of primordial germ cells

150. The best short hand notation describing this karyotype is

    (A) 46, XY
    (B) 47, XX, +18
    (C) 47, XY, +21
    (D) 47, XXY
    (E) 45, X

151. The most appropriate diagnosis of a child with this karyotype is

  (A) Down syndrome
  (B) Cri-du-chat syndrome
  (C) Marfan syndrome
  (D) Klinefelter syndrome
  (E) Turner syndrome

152. All of the following symptoms and characteristics are associated with this syndrome EXCEPT:

  (A) shortened life expectancy
  (B) cardiovascular anomalies
  (C) short stature
  (D) normal mental function
  (E) oblique palpebral fissures

153. The most predictive maternal characteristic for occurrence of this congenital birth defect is

  (A) alcoholism
  (B) diabetes mellitus
  (C) obesity
  (D) hypertension
  (E) age greater then 40 years

ANSWERS AND TUTORIAL ON ITEMS 149-153

The answers are: **149-D, 150-C, 151-A, 152-D, 153-E.** This is the karyotype of a male child with primary non-disjunctional Down syndrome (47, XY, +21). During synapsis of the first meiotic metaphase, homologous chromosomes pair. If this pairing leads to failure of separation of chromosomes, gametes are produced with either an extra copy of some particular chromosome or the lack of that copy of the chromosome. A human female is born with all of her oocytes arrested in the first meiotic metaphase. Meiosis I is not completed in the human female until just before ovulation. Consequently, synapsis can last for many years. Increasing maternal age leads to higher incidence of non-disjunctional chromosomal anomalies including Down syndrome. When non-disjunction occurs during oogenesis, the ova contain either two copies of chromosome 21 or no copies of chromosome 21, instead of the normal single copy of chromosome 21. When an ovum containing two copies of chromosome 21 is fertilized by a haploid sperm, three copies (2 maternal and one paternal) are found in the zygote and subsequent developmental stages resulting in Down Syndrome. The common symptoms of Down syndrome are short stature, straight hair, protruding tongue, oblique palpebral fissures and cardiovascular anomalies, e.g. ventricular septal defects. Children with Down syndrome have a shortened life expectancy and reduced mental function with an IQ around 60.

56

Examine the karyotype below and then answer the questions.

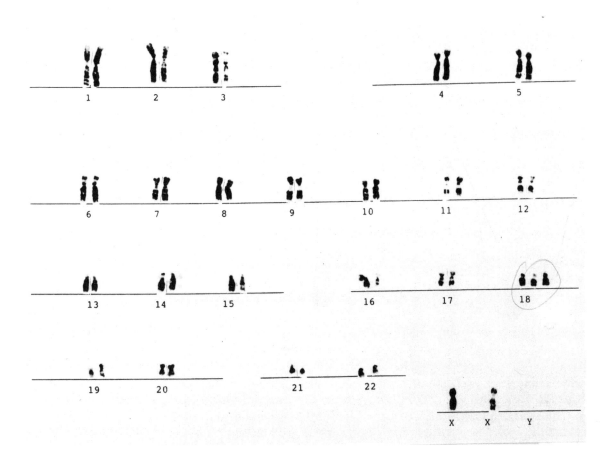

154. The developmental process whose disruption leads to the cause of this karyotype is

    (A) mitosis
    (B) synapsis
    (C) differentiation of primordial germ cells
    (D) meiotic disjunction
    (E) migration of primordial germ cells

155. The best shorthand notation describing this karyotype is

    (A) 46, XY
    (B) 47, XX, +18
    (C) 47, XY, +21
    (D) 47, XXY
    (E) 45, X

156. The most appropriate diagnosis of a child with this karyotype is

  (A) Down syndrome
  (B) Edwards syndrome
  (C) Marfan syndrome
  (D) Klinefelter syndrome
  (E) Turner syndrome

157. Which of the following symptoms is most characteristic of this aneuploid disorder?

  (A) Death soon after birth
  (B) Cardiovascular anomalies
  (C) Cleft palate
  (D) Normal mental function
  (E) Prominent occipital region of skull

158. The most predictive maternal characteristic for occurrence of this congenital birth defect is

  (A) age greater then 40 years
  (B) cocaine abuse
  (C) hypertension
  (D) high fever during pregnancy
  (E) alcoholism

ANSWERS AND TUTORIAL ON ITEMS 154-158

The answers are: **154-D, 155-B, 156-B, 157-E, 158-A.** This is the karyotype of a female child with primary non-disjunctional trisomy 18 (Edwards syndrome). This congenital anomaly occurs in approximately 1/8,000 live-born children. The appropriate shorthand notation designating this kind of aneuploidy is 47, XX, +18. During synapsis of the first meiotic metaphase, homologous chromosomes pair. If this pairing leads to failure of separation of chromosomes, gametes are produced with either an extra copy of some particular chromosome or lacking that copy of the chromosome. A human female is born with all of her oocytes arrested in the first meiotic metaphase. Meiosis I is not completed in the human female until just before ovulation. Consequently, synapsis can last for many years. Increasing maternal age leads to higher incidence of non-disjunctional chromosomal anomalies including Edwards syndrome. When non-disjunction occurs during oogenesis, the ova contain either two copies of chromosome 18 or no copies of chromosome 18 instead of the normal single copy of chromosome 18. When an ovum containing two copies of chromosome 18 is fertilized by a haploid sperm, three copies (2 maternal and one paternal) are found in the zygote and subsequent developmental stages resulting in trisomy 18 (Edwards syndrome). The common symptoms peculiar to trisomy 18 are microcephaly, prominent occiput, low-set pointed ears, micrognathia, overlapping fingers and rounded feet with a large calcaneus. Cardiovascular and renal defects are also commonly found in Edwards syndrome but are found in other conditions as well, e.g. cardiovascular defects are also common in Down syndrome. Children with trisomy 18 usually die soon after birth.

Items 159-163

Examine the karyotype below and then answer the questions.

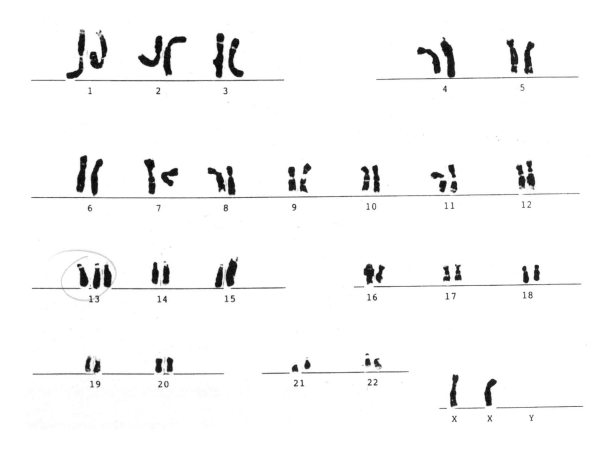

159. The developmental process whose disruption leads to the cause of this karyotype is

    (A) meiotic disjunction
    (B) block to polyspermy
    (C) polar body formation
    (D) crossing over
    (E) zona pellucida formation

160. The best shorthand notation describing this karyotype is

    (A) 46, XX
    (B) 47, XY, +18
    (C) 47, XX, +13
    (D) 48, XXXY
    (E) 46, XY

161. The most appropriate diagnosis of a child with this karyotype is

    (A) Down syndrome
    (B) Edwards syndrome
    (C) Patau syndrome
    (D) Fetal alcoholism syndrome
    (E) Turner syndrome

162. Which of the following symptoms is most characteristic of this aneuploid disorder?

    (A) Death soon after birth
    (B) Cardiovascular anomalies
    (C) Cleft palate
    (D) Normal mental function
    (E) Prominent occipital region of skull

163. The most predictive maternal characteristic for occurrence of this congenital birth defect is

    (A) heroin abuse
    (B) age greater then 40 years
    (C) pelvic inflammatory disease
    (D) eclampsia
    (E) cytomegalovirus infection

ANSWERS AND TUTORIAL ON ITEMS 159-163

The answers are: **159-A, 160-C, 161-C, 162-C, 163-B**. This is the karyotype of a female child with primary non-disjunctional trisomy 13 (Patau syndrome). This congenital anomaly occurs in approximately 1/20,000 live-born children. The appropriate shorthand notation designating this kind of aneuploidy is 47, XX, +13. During synapsis of the first meiotic metaphase, homologous chromosomes pair. If this pairing leads to failure of separation of chromosomes, gametes are produced with either an extra copy of some particular chromosome or lacking that copy of the chromosome. A human female is born with all of her oocytes arrested in the first meiotic metaphase. Meiosis I is not completed in the human female until just before ovulation. Consequently, synapsis can last for many years. Increasing maternal age leads to higher incidence of non-disjunctional chromosomal anomalies including Patau syndrome. When non-disjunction occurs during oogenesis, the ova contain either two copies of chromosome 13 or no copies of chromosome 13 instead of the normal single copy of chromosome 13. When an ovum containing two copies of chromosome 13 is fertilized by a haploid sperm, three copies (2 maternal and one paternal) are found in the zygote and subsequent developmental stages resulting in trisomy 13 (Patau syndrome). The common symptoms peculiar to trisomy 13 are holoprosencephaly, microphthalmia, anophthalmia, cleft lip, cleft palate and polydactyly. Congenital heart defects are found in Patau syndrome but are found in other conditions as well, e.g. in Down and Edwards syndrome. Children with trisomy 13 usually die soon after birth.

60

Examine the karyotype below and then answer the questions.

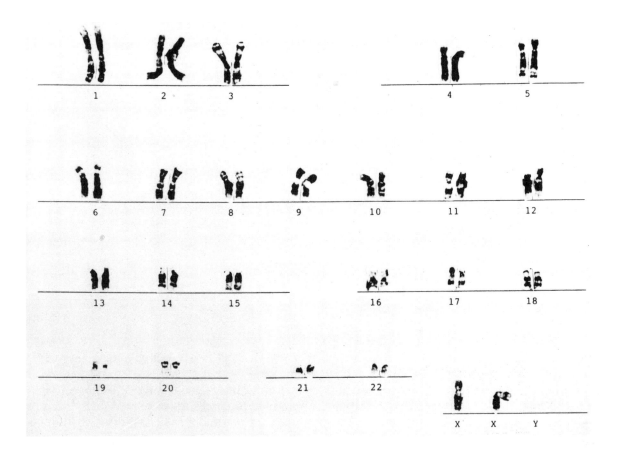

164. Which prenatal diagnostic technique is most commonly used in karyotype analysis?

    (A) Ultrasound
    (B) Amniocentesis
    (C) Fetal heart monitoring
    (D) Measurement of maternal serum α-fetoprotein
    (E) Urinalysis

165. Which drug is most commonly used in karyotype analysis?

    (A) Steroids
    (B) ß-blockers
    (C) Colchicine
    (D) ACTH
    (E) Prolactin

166. Which shorthand notation is most appropriate for describing the karyotype?

    (A) 46, XX
    (B) 47, XXY
    (C) 46, XY
    (D) 45, X
    (E) 45, Y

167. Which statement best characterizes this karyotype?

    (A) Aneuploidy with numerical anomalies in sex chromosomes
    (B) Aneuploidy with numerical anomalies in autosomes
    (C) Euploidy
    (D) Monosomy
    (E) Trisomy

168. Which of the following disorders would be associated with this karyotype?

    (A) Testicular feminization syndrome
    (B) Klinefelter syndrome
    (C) Sickle cell anemia
    (D) Turner syndrome
    (E) Down syndrome

ANSWERS AND TUTORIAL ON ITEMS 164-168

The answers are: **164-B, 165-C, 166-A, 167-C, 168-C.** This is an example of a normal female karyotype. Fetal tissue samples must be gathered for karyotype analysis. In prenatal diagnosis, fetal cells can be gathered by amniocentesis or chorionic villus sampling. Following collection of fetal cells, they can be grown in tissue culture. Treatment of cultures of cells with the drug colchicine results in the arrest of cell division at metaphase. Chromosomes on the metaphase plate can then be spread, stained, photographed and mounted in a karyotype. This karyotype is a normal female karyotype described by the shorthand notation 46, XX. This is a euploid karyotype. Aneuploid karyotypes have numerical anomalies so that there are more or less than 46 chromosomes. Trisomy and monosomy are examples of aneuploid karyotypes. Klinefelter syndrome (47, XXY) is an example of sex chromosome aneuploidy. Trisomy 21 (Down syndrome), 47, XX, +21 in a female is an example of an autosomal aneuploidy. Testicular feminization syndrome is due to a lack of the testosterone receptor and would be associated with a normal male karyotype (46, XY). Turner syndrome is due to a lack of a second X chromosome (45, X). Sickle cell anemia is due to a point mutation in the hemoglobin gene and would be associated with either a normal male or normal female karyotype.

Items 169-172

A fetus is delivered at 32 weeks since the last menstrual period. At birth, the fetus weighs 1500 gm but otherwise appears normal. Soon after birth, however, the fetus becomes cyanotic and breathes with a grunting noise. Chest X-rays reveal dense lungs with significant atelectasis.

169. The most likely diagnosis associated with these symptoms is

    (A) congenital diaphragmatic hernia
    (B) coarctation of the aorta
    (C) tetralogy of Fallot
    (D) renal agenesis
    (E) respiratory distress syndrome

170. The most significant predictive prenatal test for this disease would be

    (A) ultrasonography
    (B) X-rays
    (C) measurement of amniotic fluid lecithin/sphingomyelin ratio
    (D) karyotype analysis
    (E) measurement of maternal serum α-fetoprotein

171. The cell type most directly involved in etiology of this condition is

    (A) type II pneumocyte
    (B) macrophage
    (C) ciliated cell
    (D) goblet cell
    (E) erythrocyte

172. In threatened premature delivery, the most useful drug class for treatment of the mother to accelerate fetal lung maturation is

    (A) antibiotic
    (B) antimetabolite
    (C) neuroleptic
    (D) steroid
    (E) diuretic

ANSWERS AND TUTORIAL ON ITEMS 169-172

The answers are: **169-E, 170-C, 171-A, 172-D.** The leading cause of death among premature infants is respiratory distress syndrome. A fetus 30 weeks from the last menstrual period is 28 weeks into a gestation. At 28 weeks gestational age, type II pneumocytes first differentiate and begin to secrete surfactant. The most abundant phospholipid in surfactant is lecithin. The sphingomyelin levels in amniotic fluid are relatively constant throughout gestation but the lecithin levels begin to rise slowly at 28 weeks. Once many type II cells differentiate, they secrete substantial quantities of surfactant which enters the amniotic fluid due to fetal respiratory movements. Due to increasing surfactant secretion, the lecithin/sphingomyelin ratio increases. Type II cell differentiation is a steroid dependent process. Treatment of women with certain steroid drugs will stimulate precocious differentiation of type II cells and thus reduce the risk of respiratory distress syndrome in premature infants.

Items 173-177

Examine the labeled scanning electron micrograph of the pharyngeal apparatus below and then match the lettered structure in the micrograph with the most appropriate description of its developmental fate or involvement in congenital abnormalities in the questions below.

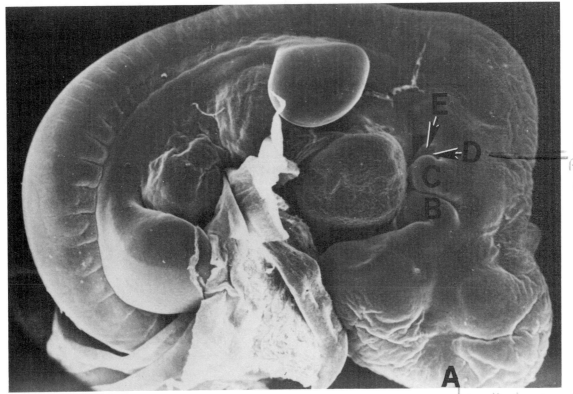

173. Developmental anomalies in this structure cause the mandibular hypoplasia seen in mandibulofacial dysostosis (Treacher-Collins syndrome).

174. This structure forms the maxilla.

175. The aortic arch in this pharyngeal arch forms the hyoid and stapedial arteries in the adult.

176. This structure forms the external auditory meatus in the adult.

177. The mandible and masseter muscles form in the mesenchyme of this structure.

ANSWERS AND TUTORIAL ON ITEMS 173-177

The answers are: **173-C, 174-B, 175-E, 176-D, 177-C.** This is a scanning electron micrograph of a human embryo of approximately 5 weeks since fertilization. The forelimb bud and hindlimb bud are both visible. The frontal prominence (A) will become the forehead. The first pharyngeal arch is divided into a maxillary process (B) and a mandibular process (C), both of which will grow and fuse in the ventral midline. The first pharyngeal cleft (D) is clearly visible between the first and second (E) pharyngeal arches. The maxillary processes will form the maxilla. The mandibular processes will form the mandible. Mandibulofacial dysostosis (first arch syndrome, Treacher Collins syndrome) is caused by an autosomal dominant mutation with a sex ratio of 1:1. This mutation causes developmental anomalies in derivatives of the first pharyngeal arch. The major clinical findings associated with this anomaly are downward slanting palpebral fissures, malar hypoplasia, microtia and conductive hearing loss due to abnormalities in the auditory ossicles and external auditory meatus. This hearing loss is often misdiagnosed as mental retardation which is not a characteristic of people with mandibulofacial dysostosis. The aortic arches of the first and second pharyngeal arches undergo extensive degenerative changes during development but persist in the adult as the maxillary artery (first arch) or hyoid and stapedial artery (second arch). The first pharyngeal cleft gives rise to the external auditory meatus while the first pharyngeal pouch forms the auditory tube. The caudal pharyngeal clefts disappear due to caudal overgrowth by the second pharyngeal arch. The second pharyngeal pouch forms the palatine tonsils while the third and fourth pouches both contribute to the thymus and the parathyroid glands. The muscles of mastication are formed in first arch mesenchyme and are innervated by the mandibular division of the Vth (trigeminal) cranial nerve. The muscles of facial expression are formed in the second arch mesenchyme and are innervated by the VIIth (facial) cranial nerve.

Treacher syndrome - slanting of palpebral fissure

Items 178-179

A mother and her two year old son come to your office. The mother complains that the child is not acting normally. After a brief period of active play, her son becomes exhausted, breathes rapidly and squats or lies down. On physical examination, you find that the child's pulse and blood pressure are normal. His oral mucosa, fingernails and toenails are cyanotic. You also detect a significant systolic murmur best heard along the left sternal border at the level of the second intercostal space. A PA (posteroanterior) chest film shows a concavity on the left border of the heart, diminished pulmonary vascularity and a rounded heart apex located slightly higher than normal above the diaphragm.

178. The most likely diagnosis of the boy's problem is

   (A) isolated ventricular septal defect
   (B) isolated atrial septal defect
   (C) isolated aortic stenosis
   (D) transposition of the great vessels
   (E) tetralogy of Fallot

179. All of the following are common anatomical features of this condition **EXCEPT:**

   (A) pulmonary stenosis
   (B) left ventricular hypertrophy
   (C) overriding aorta
   (D) ventricular septal defect
   (E) patent ductus arteriosus

66

ANSWERS AND TUTORIAL ON ITEMS 178-179

The answers are: **178-E, 179-B**. This child has the tetralogy of Fallot. The mild cyanosis and dyspnea are caused by the poor vascular perfusion of the lungs due to the pulmonary artery stenosis and ventricular septal defect with large right to left shunt. Abnormally high pressure in the right ventricle due to the overriding aorta causes right ventricular hypertrophy and is largely responsible for the radiological findings. The systolic murmur is due to turbulence of blood flow through the right ventricular outflow tract. The primitive heart, prior to ventricular septation, has a single large ventricle which leads into a large ventricular outflow tract called the conus cordis. The conus cordis leads into the truncus arteriosus which in turn supplies blood to the aortic sac and aortic arches. The tetralogy of Fallot is the most frequently seen abnormality caused by unequal division of the conus cordis and truncus arteriosus. This defect is caused by an anterior displacement of the truncoconal septum, a flap of tissue that forms in the conus cordis and truncus arteriosus to divided the ventricular outflow tracts. Prior to birth, the ductus arteriosus, a derivative of the left distal 6th aortic arch, provides a shunt between the pulmonary and the systemic circulation. Soon after birth, the ductus arteriosus closes and degenerates, leaving behind a ligamentum arteriosum. In many cases of tetralogy of Fallot, there is also a persistent ductus arteriosus.

Items 180-181

During a routine obstetrical examination, you find that your patient has an abnormally small abdominal girth for her 30 week pregnancy. Ultrasonographic examination reveals a normal fetus, a posterior fundal placenta and oligohydramnios.

180. The decreased volume of amniotic fluid (oligohydramnios) could be caused by which of the following?

    (A) Renal agenesis
    (B) Bladder exstrophy
    (C) Anencephaly
    (D) Tracheoesophageal fistula
    (E) Duodenal torsion

181. The most likely fundamental developmental defect leading to this clinical case is

    (A) failed ureteric bud induction
    (B) failed cloacal membrane septation
    (C) failed neural tube closure
    (D) failed secondary canalization
    (E) failed mid-gut loop rotation

The answers are: **180-A, 181-A**. Complete renal agenesis most likely results from a failure of development of the metanephric (definitive) kidney. The fundamental process leading to formation of the metanephros is a reciprocal inductive interaction between the ureteric bud (a branch of the mesonephric duct) and the metanephric blastema (the caudal portion of the urogenital ridge). The ureteric bud grows into the metanephric blastema and is induced to branch many times leading to the formation of the ureters, major calyces, minor calyces and collecting ducts of the definitive kidney. The ureteric bud induces the formation of nephrons in the metanephric blastema. Derivatives of the metanephric blastema include the renal glomerulus and other blood vessels, connective tissue, Bowman's capsule, proximal convoluted tubule, loop of Henle and distal convoluted tubule.

The fetus normally produces a small amount of hypotonic urine and urinates it into the amniotic fluid. Subsequently, the fetus swallows the amniotic fluid. Renal agenesis results in no urine formation and therefore a reduction in volume of amniotic fluid (oligohydramnios). Anencephalic fetuses have no swallowing function but do produce urine, leading to the excessive volume of amniotic fluid (polyhydramnios). Urine production is normal in bladder exstrophy which does not become apparent as a clinical problem until after birth. Blockage of the gastrointestinal tract would also lead to polyhydramnios.

A 42 year-old pregnant woman has chorionic villus sampling performed for cytogenetic analysis of her fetus. The scanning electron micrograph below is prepared from this sample. Examine this micrograph and then answer the questions concerning the developing placenta.

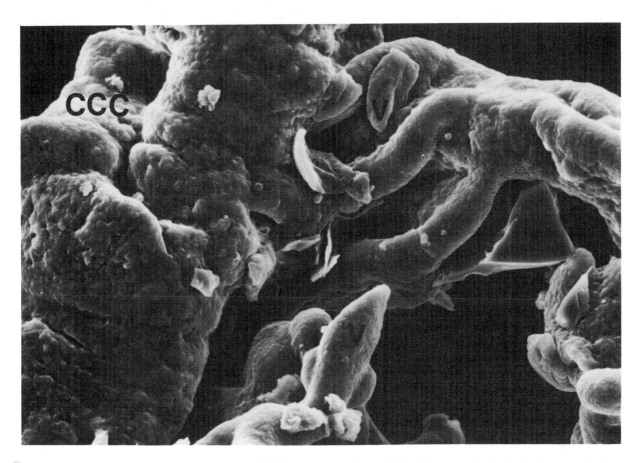

182. The cytotrophoblastic cell columns (CCC) have all of the following morphological characteristics **EXCEPT**:

    (A) fetal blood vessels
    (B) maternal blood vessels
    (C) connective tissue
    (D) cytotrophoblast
    (E) syncytiotrophoblast

183. The villi (V) have all of the following morphological characteristics **EXCEPT**:

    (A) a continuous layer of cytotrophoblastic cells
    (B) a continuous syncytiotrophoblast
    (C) a connective tissue core
    (D) decidual cells
    (E) they are bathed in maternal blood

ANSWERS AND TUTORIAL ON ITEMS 182-183

The answers are: **182-B, 183-D**. This specimen is a scanning electron micrograph of a sample of chorionic villi. Chorionic villus sampling is gradually replacing amniocentesis as a technique for gathering fetal tissue for cytogenetic analysis. This technique has several advantages. First, it can be performed earlier than amniocentesis and thus can give diagnosis of cytogenetic or biochemical abnormalities at an earlier stage in pregnancy. It also yields samples of rapidly growing fetal tissue which can be grown into substantial numbers of cells for karyotype or biochemical analysis more easily and more rapidly than fetal tissue samples collected by amniocentesis. Cytotrophoblastic cell columns (CCC) connect the chorionic plate and the basal plate of the developing placenta. They consist of columns of fetal connective tissue surrounding fetal blood vessels. The cytotrophoblastic cell columns are covered by the layer of cytotrophoblastic cells and a layer of syncytiotrophoblast. They contain no maternal blood vessels. Chorionic villi (V) project from the cytotrophoblastic cell columns into the intervillous spaces. Villi are bathed in maternal blood that enters the intervillous spaces through branches of the uterine artery that penetrate the chorionic plate. Villi also contain fetal blood vessels surrounded by fetal connective tissue. The villi are coated by a layer of cytotrophoblast and syncytiotrophoblast. Maternal decidual tissue is found in the placenta but not in the villi proper.

Items 184-186

A 37 year-old 148 lb woman is found to be carrying monoamnionic twins. In spite of complete continuous bed rest from 28 weeks after conception, she is found to have ruptured membranes at 34 weeks. Two days after rupture of membranes, maternal temperature is 103° F. and her white blood count is 20,000/mm³ with a predominance of neutrophils. Your diagnosis is chorioamnionitis.

184. The chief morphological change in the fetal lymph nodes as a result of this intrauterine infection would be

  (A) ingression
  (B) increase in T-cells
  (C) infiltration by neutrophils
  (D) appearance of eosinophils
  (E) appearance of germinal centers

185. These morphological changes are a reflection of which immunological process?

  (A) T-cell proliferation
  (B) Activation of memory B-cells
  (C) Delayed hypersensitivity
  (D) Plasma cell phagocytosis
  (E) B-cell activation and differentiation

186. Which mechanism is most important for passive immunization of the fetus against antigenic exposure of the mother prior to birth?

  (A) Placental transport of maternal IgGs
  (B) Synthesis of secretory IgAs
  (C) Placental transport of maternal IgMs
  (D) Colostrum formation
  (E) Placental transport of IgEs

ANSWERS AND TUTORIAL ON ITEMS 184-186

The answers are **184-E, 185-E, 186-A.** Premature rupture of fetal membranes often leads to chorioamnionitis with elevated temperature and elevated white blood cell count. The normal white blood cell count is approximately 5,000/mm³. Under normal circumstances, the fetus develops in a sterile environment and is therefore not exposed to bacterial antigens. Consequently, fetal lymph nodes lack germinal centers. At birth, the fetus is suddenly exposed to an entire new population of bacterial antigens. This potential problem is dealt with in three ways. First, during gestation, the fetus is passively immunized against bacterial antigens by transplacental transport of maternal IgGs. Higher molecular weight IgM is not transported across the placenta. Second, if the child is breast fed, colostrum contains high concentrations of secretory IgAs that bind to enteric bacteria and prevent their adhesion to the gastrointestinal epithelium. Third, once the fetus becomes exposed to bacterial antigens, his or her own lymph nodes (if antigens are lymph-born) and spleen (if the antigens are blood-born) become activated. Germinal centers represent areas where B-lymphocytes, in conjunction with helper T-cells and antigen-presenting macrophages are becoming activated. This activation involves decreasing nuclear condensation and increasing cytoplasm to produce the intracellular machinery for IgG synthesis. Once stimulated, B-cells differentiate into immunoglobulin secreting plasma cells. Germinal centers also contain large numbers of macrophages.

Items 187-190

In a second pregnancy, a 25 year old woman delivers a hydropic fetus with severe edema, jaundice and hepatosplenomegaly. Her blood type is O negative and her husband's blood type is O positive.

187. The blood type of the fetus is

    (A) A negative
    (B) A positive
    (C) O negative
    (D) O positive
    (E) B positive

188. All of the following statements concerning the fetal red blood cells are correct **EXCEPT**:

    (A) they lack Rh antigen
    (B) they are destroyed in the fetal spleen
    (C) they are produced in the fetal liver
    (D) they are produced in fetal bone marrow
    (E) they are coated with maternally derived IgG

189. The most appropriate description of the cause of fetal hepatosplenomegaly is

    (A) maternal IgGs cause stimulation of Kupffer cells
    (B) liver and spleen enlarge to facilitate removal of hemoglobin degradation products
    (C) these organs are active in destruction and production of fetal red blood cells
    (D) fetal blood volume increased so these organs have more blood in them than usual
    (E) spleen and liver show hypertrophy to substitute for compromised placental function

190. The primary cause of Rh isoimmunization is

    (A) leakage of fetal RBCs into maternal blood supply
    (B) accumulation of fetal IgGs at placenta
    (C) failure of transplacental transport of maternal IgGs to fetus
    (D) immunization of mother against paternal spermatozoa
    (E) abnormal hepatic morphogenesis

ANSWERS AND TUTORIAL ON ITEMS 187-190

The answers are: **187-D, 188-A, 189-C, 190-A.** Rh isoimmunization occurs in fetuses with Rh-mothers and Rh + fathers. The Rh antigen is controlled by a dominant gene and therefore the fetus of an Rh- mother and Rh + father would be Rh + . Rh isoimmunization is more severe with each pregnancy because the maternal immune system has a memory for Rh isoimmunization. Normally, the placental trophoblast serves as a barrier between fetal and maternal blood, although microvascular accidents allowing mixture of fetal and maternal blood are inevitable. When fetal RBCs leak into the maternal blood supply, they are recognized as foreign antigens by the maternal immune system. Maternal IgGs are produced against the fetal RBCs and these are transported functionally intact across the placenta into the fetal circulatory system. Here, these IgGs bind to fetal RBCs and cause their destruction in the fetal liver and spleen. The resulting anemia stimulates fetal erythropoiesis in the bone marrow, liver and spleen. Consequently, enlargement of the liver and spleen occurs. Excessive destruction of fetal RBCs results in accumulation of bile pigments and thus jaundice.

Items 191-192

You are a Public Health Service pediatrician working on an Indian Reservation in Montana. An obese woman with hepatomegaly and facial spider angiomas visits your clinic with her 4 year old son. The mother admits to drinking whiskey on a regular basis but denies alcohol consumption during her pregnancy. The child is short for his age and shows significant mid-facial hypoplasia , including ocular ptosis and a flat nasal bridge. The child seems lethargic and there is a significant deficit in his social interactions.

191. The most likely diagnosis of the maternal-child disease is

    (A) healthy-Down syndrome
    (B) healthy-Treacher-Collins syndrome
    (C) alcoholism-fetal alcohol syndrome
    (D) alcoholism-vitamin B_{12} deficiency
    (E) alcoholism-pellagra

192. The most likely defective developmental process that lead to the child's symptoms is

    (A) neural crest cell migration
    (B) bone formation
    (C) ocular induction
    (D) cartilage proliferation
    (E) endochondral ossification

ANSWERS AND TUTORIAL ON ITEMS 191-192

The answers are: **191-C, 192-A.** There is a high probability that the mother is an alcoholic. Her obesity, hepatomegaly and spider angiomas are highly suggestive of alcoholism. Her denial of alcohol consumption during pregnancy is suspect. Alcoholism is especially common among American Indians. The child shows many of the features of fetal alcohol syndrome (FAS) including short stature, facial abnormalities and mild mental retardation. Alcohol and acetaldehyde are known teratogens. They cross the placenta rapidly and persist in the fetus long after being cleared from the maternal system. The fetal hepatic detoxification functions for alcohol are poorly developed. Experimental studies in mammalian embryos suggest that one of the primary causative events in FAS teratogenesis is inhibition of cell migration. Neural crest cells migrate into the pharyngeal apparatus where they participate in morphogenesis of facial bones. The mid-facial hypoplasia characteristic of FAS children is most likely due to defects in neural crest cell migration.

Items 193-194

A new born infant fails to defecate for 3 days after birth in spite of normal feeding without excessive vomiting. Rectal examination reveals a normal rectum but a fecal mass retained in the colon. There is an explosive release of feces following cessation of rectal examination. Barium enema and x-rays reveals a narrowed distal segment of the colon and a dilated proximal segment.

193. The most likely diagnosis of this condition is

    (A) sprue
    (B) colic
    (C) colonic aganglionosis
    (D) imperforate anus
    (E) gastric atresia

194. The most likely developmental abnormality leading to this condition is

    (A) gluten allergy
    (B) abnormal gastric rotation
    (C) abnormal neural crest cell migration
    (D) failure of proctodeal membrane degeneration
    (E) situs inversus

The answers are: **193-C, 194-C**. This patient is suffering from Hirschsprung's disease or colonic aganglionosis. Biopsy of the distal colon would reveal that the parasympathetic ganglia and other neuronal elements of the myenteric plexus are absent from the narrowed portion of the colon. Ganglia are present in the distended portion of the colon. The neurons of the myenteric plexus are derived from the neural crest and the congenital anomaly is due to abnormal neural crest cell migration to the colon. Neural crest cells also contribute to dorsal root ganglia; form the adrenal medulla; form Schwann cells, myelinating cells in the peripheral nervous system; form meninges; form melanocytes and form odontoblasts, the dentin-secreting cells of tooth germs.

Items 195-196

Contrast media can be injected into blood vessels to reveal by x-ray the blood flow from one vessel to another. In the hemochorial human placenta, a contrast dye is injected into either the uterine artery or the umbilical vein and 10 sec later the dye is located by radiography.

195. The vascular space most intensely labeled immediately after injection of contrast medium into the uterine artery is

   (A) the umbilical artery
   (B) the umbilical vein
   (C) capillaries in the chorionic villi
   (D) the intervillous space
   (E) the uterine vein

196. The vascular space most intensely labeled immediately after injection of contrast medium into the umbilical vein is the

   (A) inferior vena cava
   (B) ductus venosus
   (C) ductus arteriosus
   (D) umbilical artery
   (E) superior vena cava

The answers are: **195-D, 196-A.** Maternal oxygenated blood is supplied to the placenta from branches of the uterine arteries. These vessels empty into the intervillous space where they carry oxygen to and remove carbon dioxide from capillaries in chorionic villi. The deoxygenated maternal blood returns to the uterine veins and eventually the maternal pulmonary circulation, where carbon dioxide is expelled and fresh oxygen is dissolved in maternal blood for return to the uterine artery. Deoxygenated fetal blood enters the placenta through the umbilical arteries. These blood vessels send branches into the chorionic villi where there is a complex anastomosing network of capillaries. In these villi, fetal blood becomes oxygenated and then drains back into the umbilical vein. The umbilical vein carrying oxygenated fetal blood empties into the fetal inferior vena cava, shunts by the liver through the ductus venosus and enters the right atrium. Before birth, this blood is then mostly shunted through the foramen ovale into the left atrium and then into the fetal systemic circulation. Some of this oxygenated blood mixes with deoxygenated blood returning to the right atrium from the superior vena cava and then enters the right ventricle where it is pumped into the pulmonary artery. Most of the blood leaving the right ventricle is shunted away from the lungs into the systemic circulation at the arch of the aorta by way of the ductus arteriosus.

Items 197-198

During a squash match, a 35 year old male experiences sharp pain in his lower back. The next day, he experiences stiffness in his back, complete lack of flexion of the vertebral column due to pain, a dull intense pain in the right buttock and numbness in the lateral surface of the right lower limb. MRI reveals a severe herniated L4-L5 intervertebral disc on the right side.

197. The structure causing the pain in the buttock and numbness in the lower limb is the

   (A) annulus fibrosus
   (B) nucleus pulposus
   (C) ligamentum flavum
   (D) vertebral arch
   (E) transverse vertebral process

198. The offending structure is derived from which embryonic structure?

   (A) Notochord
   (B) Intermediate mesoderm
   (C) Dermatome
   (D) Myotome
   (E) Sclerotome

ANSWERS AND TUTORIAL ON ITEMS 197-198

The answers are: **197-B, 198-A**. Ruptured intervertebral discs are caused by a traumatic tearing of the annulus fibrosus followed by expulsion of the plastic nucleus pulposus so that it impinges upon the nerve roots, causing pain, numbness and possible loss of motor function. Lower back pain is particularly common in humans because we have a vertebral column better suited for quadrupedal locomotion in spite of the fact that we are bipedal. Our erect posture places large compressive forces on the lumbar intervertebral discs. The notochord is a mesodermally derived structure that runs along much of the length of the embryonic body beneath the neural tube. As the vertebral column forms from the sclerotomes, the notochordal component of the vertebral bodies degenerates and disappears. The notochordal component of the intervertebral discs, however, persists in the adult as the nucleus pulposus.

Examine the photomicrograph below and then answer the questions concerning this photomicrograph. It is a histological preparation of a developing human embryo sectioned through the thoracic level of the body and includes the developing heart (H).

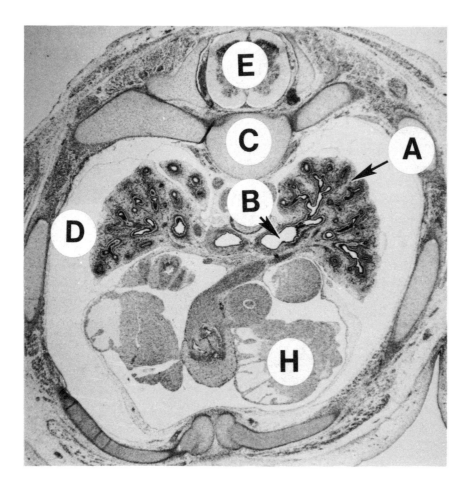

199. The embryonic rudiment labeled A gives rise to all of the following structures **EXCEPT**:

    (A) blood vessels
    (B) connective tissue
    (C) pulmonary macrophages
    (D) visceral pleura
    (E) lymphatic vessels

200. The embryonic rudiment labeled B gives rise to all of the following structures **EXCEPT**:

    (A) ciliated cells
    (B) goblet cells
    (C) brush cells
    (D) basal cells
    (E) chondrocytes

201. The embryonic rudiment labeled C gives rise to which adult structure?

    (A) Neural arch
    (B) Transverse process
    (C) Spinous process
    (D) Nucleus pulposus
    (E) Vertebral body

202. The space labeled D is derived from the _____ and becomes the _____.

    (A) extraembryonic coelom-allantois
    (B) intraembryonic coelom-peritoneal cavity
    (C) extraembryonic coelom-diaphragm
    (D) intraembryonic coelom-pleural cavity
    (E) septum transversum-diaphragm

203. What is the approximate age of this specimen from time of fertilization?

    (A) 1-3 weeks
    (B) 4-5 weeks
    (C) 6-8 weeks
    (D) 10-12 weeks
    (E) 16-18 weeks

204. The underdeveloped organ labeled by A and B will become capable of supporting extrauterine life in approximately

    (A) 14 weeks
    (B) 16 weeks
    (C) 18 weeks
    (D) 20 weeks
    (E) greater than 20 weeks

ANSWERS AND TUTORIAL ON ITEMS 199-204

The answers are: **199-C, 200-E, 201-E, 202-D, 203-C, 204-E**. This is a histological section through the developing heart and lungs taken at 6-8 weeks of development. The developing lungs are in the glandular phase of histogenesis which extends from 4 weeks to 15 weeks. During this phase, the trachea, bronchi and bronchioles form. The canalicular phase covers 16 weeks to 24 weeks and is a period of formation of terminal bronchioles, respiratory bronchioles and blood vessels. The saccular stage extends from 25 weeks to 38 weeks and is a period when many alveoli form. Type II cells, producing surfactant, first appear around 28 weeks of development. When large numbers of these cells differentiate, the lungs become functionally mature and the fetus becomes viable in the extrauterine environment. Thus, these lungs will not be functionally mature for more than 20 weeks after the period illustrated in the photomicrograph. The lungs form from two separate rudiments: the splanchnic mesoderm (A) and the lung buds (B). The respiratory diverticulum arises as a branch from the foregut at 4 weeks. The respiratory diverticulum grows caudally and branches into left and right lung buds. The lung buds continue to branch and grow into the splanchnic mesoderm. The epithelial components of the respiratory system lining the airways are derived from the respiratory diverticulum. Thus, the epithelial linings of the trachea, bronchi, bronchioles and alveoli are all derived from the respiratory diverticulum. The blood and lymphatic vessels, connective tissue, smooth muscle and cartilage in the remainder of the respiratory system are derived from splanchnic mesoderm. Pulmonary macrophages, like all other components of the mononuclear phagocyte system, are derived from bone marrow stem cells. C is a sclerotome-derived hyaline cartilage destined to become a vertebral body by endochondral ossification. The intraembryonic coelom is the precursor of the pleural cavities (D), pericardial cavity and peritoneal cavity. The growth of the pleuropericardial membranes and the diaphragm lead to the division of the intraembryonic coelom into these four separate serous cavities. E is the neural tube.

CHAPTER III

GROSS ANATOMY

Items 205-208

A 21 year-old man falls through a glass window and suffers a deep gash in the posterolateral aspect of the left side of the neck. Examination reveals that the level of the left shoulder is lower than that of the right shoulder and that the patient has difficulty in shrugging the left shoulder against resistance. The examiner also finds that there is no increase in the tonicity of the muscle mass underlying the medial part of the upper border of the left shoulder when the patient attempts to shrug the left shoulder.

205. Which nerve or part of the brachial plexus has been cut?

    (A) Axillary nerve
    (B) Spinal root of the accessory nerve
    (C) Dorsal scapular nerve
    (D) Suprascapular nerve
    (E) Upper trunk of the brachial plexus

206. Which of the following active movements will also prove difficult for the patient?

    (A) Flexing the left arm 60 degrees
    (B) Extending the left arm 60 degrees
    (C) Abducting the left arm more than 90 degrees
    (D) Internally rotating the left arm
    (E) Externally rotating the left arm

207. A deep gash in the posterolateral aspect of the neck may result in any of the following sensory deficits **EXCEPT:**

    (A) a loss of sensation to touch in the skin of the anterior aspect of the neck
    (B) a loss of sensation to pain in the skin overlying the clavicle
    (C) a loss of sensation to temperature in the skin overlying the angle of the mandible
    (D) a loss of sensation to touch in the skin of the lobule of the ear
    (E) a loss of sensation to temperature in the skin of the chin

208. A deep gash in the posterior triangle of the neck can cut the

    (A) common carotid artery
    (B) external carotid artery
    (C) internal carotid artery
    (D) external jugular vein
    (E) internal jugular vein

ANSWERS AND TUTORIAL FOR ITEMS 205-208

The answers are: **205-B; 206-C; 207-E; 208-D.** Flaccidity of the muscle mass underlying the medial part of the upper border of the shoulder indicates trapezius palsy. In this case, trapezius palsy suggests a cut through the spinal root of the accessory nerve in the posterior triangle of the neck. A deep gash in the posterolateral aspect of the neck may cut nerves and blood vessels within the posterior triangle of the neck, such as the spinal root of the accessory nerve, branches of the cervical plexus, the supraclavicular parts of the brachial plexus and the external jugular vein. The posterior triangle of the neck is bordered anteriorly by the posterior border of sternocleidomastoid (S), posteriorly by the anterior border of trapezius (T) and inferiorly by the middle third of the clavicle (C) (Fig. 1).

The spinal root of the accessory nerve arises from fibers that emerge from the five uppermost cervical spinal cord segments (C1-C5). The fibers form a trunk within the vertebral canal that first ascends into the cranial cavity (by passing through the foramen magnum) before descending into the neck (by passing through the jugular foramen). As the spinal root of the accessory nerve descends through the neck, it first pierces (and innervates) sternocleidomastoid, then crosses the posterior triangle of the neck and finally passes deep to trapezius (in route to its innervation of trapezius). The spinal root of the accessory nerve (N) is especially susceptible to injury along its descent across the posterior triangle of the neck, for here it lies covered only by skin, superficial fascia and the investing layer of deep cervical fascia (Fig. 1).

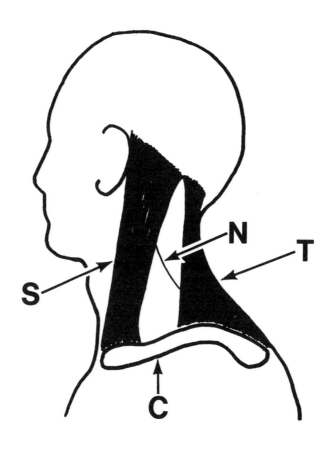

Fig. 1

Trapezius is the chief muscle that supports the scapula and clavicle from the axial skeleton. It is the muscle chiefly responsible for the capacity to raise the shoulder. Trapezius and serratus anterior act together as the prime movers for abducting the arm from 60 to 150 degrees. Consequently, loss of trapezius's innervation results in a lowering of the shoulder and weakness in either shrugging the shoulder or abducting the arm above the shoulder. Isolated trapezius palsy also results in flaring of the vertebral border and inferior angle of the scapula. Abduction of the arm against resistance accentuates the flaring, but flexion of the arm minimizes it.

The cutaneous branches of the cervical plexus all emerge around the posterior border of sternocleidomastoid in route to their areas of cutaneous innervation. The transverse cervical nerve (from C2 and C3) provides sensory innervation (pain, temperature and touch sensation) for almost all the skin overlying the anterior triangle of the neck (region TC in Fig. 2). The supraclavicular nerves (from C3 and C4) provide sensory innervation for the skin overlying the top of the shoulder (in particular that overlying the clavicle) (region SC in Fig. 2). The great auricular nerve (from C2 and C3) provides sensory innervation for the skin overlying the angle of the mandible, the lower part of the parotid gland, the mastoid process and almost all of the auricle (including the lobule) (region GA in Fig. 2). The lesser occipital nerve (from C2 and C3) provides sensory innervation for the strip of the scalp immediately posterior to the auricle (region LO in Fig. 2). The mandibular division of the trigeminal nerve provides sensory innervation for the skin of the chin.

The external jugular vein descends across the lower part of the posterior triangle of the neck before passing through the investing layer of deep cervical fascia to join the subclavian vein. The investing layer of deep cervical fascia is attached to the outer surface of the external jugular vein at the site where the vein pierces the fascia. This fascial attachment tends to keep the vein open. Consequently, if the external jugular vein is cut anywhere above its attachment to the investing layer of deep cervical fascia in the posterior triangle of the neck, there is the risk that a pulmonary air embolism will occur as a result of air being sucked into the vein during inspiration.

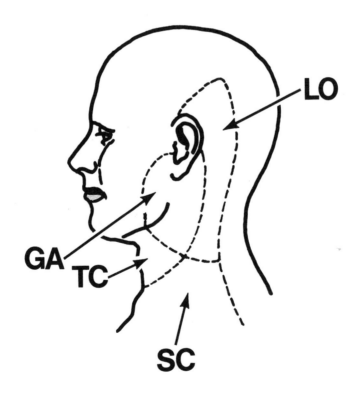

Fig. 2

Items 209-211

An 18 year-old man suffers a painful injury to the right shoulder while playing football. An examiner learns that the injury was the result of a downward blow on the point of the shoulder. Visual inspection shows that the acromion process of the scapula lies anteroinferior to the lateral end of the clavicle. Palpation reveals tenderness in the region between the acromion process and the lateral end of the clavicle. The patient reports pain when abducting the right arm up to or above the level of the shoulder. Anteroposterior (AP) radiographs of the patient's shoulders show that the acromioclavicular and coracoclavicular spaces in the right shoulder are each more than 100% wider than the corresponding spaces in the left shoulder.

209. Which bony structure contributes to the shape of the point of the shoulder?

    (A) Acromion process of the scapula
    (B) Coracoid process of the scapula
    (C) Spine of the scapula
    (D) Lateral end of the clavicle
    (E) Head of the humerus

210. Which of the following ligaments is significantly ruptured in the patient's right shoulder?

    (A) Costoclavicular ligament
    (B) Coracoacromial ligament
    (C) Coracohumeral ligament
    (D) Coracoclavicular ligament
    (E) Suprascapular ligament

211. Which injury is indicated by the history, physical exam and radiographs?

    (A) Acromioclavicular joint sprain
    (B) Acromioclavicular joint subluxation
    (C) Acromioclavicular joint dislocation
    (D) Shoulder joint dislocation
    (E) Sternoclavicular joint dislocation

ANSWERS AND TUTORIAL FOR ITEMS 209-211

The answers are: **209-A; 210-D; 211-C.** The point of the shoulder is the region at the lateral limit of the shoulder. The acromion process of the scapula shapes the point of the shoulder. Inferiorly directed blows on the point of the shoulder strain the fibrous structures that suspend the scapula from the clavicle, in particular, the capsule of the acromioclavicular joint (AC) and the coracoclavicular ligament (CL) (Fig. 3). In this case, the anteroinferior displacement of the acromion process indicates dislocation of the acromioclavicular joint and the radiographs confirm the diagnosis.

Injuries of the acromioclavicular joint are called shoulder separations. Tenderness over the acromioclavicular joint is common. Abduction of the arm beyond 90 degrees is frequently painful. The severity of a shoulder separation is assessed by comparing the widths of the acromioclavicular and coracoclavicular spaces in an AP radiograph of the injured shoulder with the widths of the corresponding spaces in an AP radiograph of the contralateral, uninjured shoulder. The acromioclavicular space is the radiolucent space between the acromion process and the lateral end of the clavicle. The space marks the location of the apposed articular cartilages in the acromioclavicular joint. The coracoclavicular space is the radiolucent space between the coracoid process of the scapula and the clavicle above. It marks the location of the coracoclavicular ligament. The coracoclavicular ligament is the principal ligamentous structure that suspends the scapula from the clavicle.

Dislocation of the acromioclavicular joint occurs when both the acromioclavicular joint capsule and the coracoclavicular ligament are significantly or completely ruptured. An AP radiograph of a dislocated acromioclavicular joint shows acromioclavicular and coracoclavicular spaces whose widths are at least 50% greater than those of the corresponding spaces in the AP radiograph of the uninjured shoulder. Inspection of the injured shoulder commonly shows the acromion process displaced anteroinferiorly to the lateral end of the clavicle.

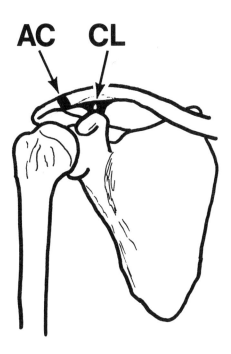

Fig. 3

Items 212-214

A 19 year-old woman awakes with a painful right shoulder the day after playing a basketball game. An examiner learns that the patient fell twice during the game, each time using the right upper limb to protect the body against impact with the floor. Visual inspection of the right shoulder does not reveal any deformity. Palpation reveals tenderness deep to the acromion process. Physical examination shows that although active abduction of the right arm from 0 to 60 degrees and from 120 to 180 degrees is painless, active or passive abduction from 60 to 120 degrees is painful.

212. Which of the following injuries is suggested by the history and physical exam?

    (A) Shoulder joint dislocation
    (B) Tear of the deltoid muscle
    (C) Fracture of the clavicle
    (D) Incomplete rupture of the supraspinatus tendon
    (E) Complete rupture of the supraspinatus tendon

213. Each of the following statements concerning arm abduction over the 60 to 120 degree arc is correct **EXCEPT:**

    (A) All of the movement is due to abduction of the arm at the shoulder joint.
    (B) Serratus anterior is a prime mover of abduction over the 60 to 120 degree arc.
    (C) Trapezius is a prime mover of abduction over the 60 to 120 degree arc.
    (D) The supraspinatus tendon is pulled proximally.
    (E) The subacromial bursa is pulled proximally.

214. Each of the following muscles contributes to the rotator cuff of the shoulder joint **EXCEPT:**

    (A) teres minor
    (B) deltoid
    (C) supraspinatus
    (D) infraspinatus
    (E) subscapularis

ANSWERS AND TUTORIAL FOR ITEMS 212-214

The answers are: **212-D; 213-A; 214-B.** A painful arc of arm abduction that extends from about 60 to 120 degrees suggests a lesion of the supraspinatus tendon of insertion or inflammation of the subacromial bursa. The history of injury to the shoulder and the presence of tenderness deep to the acromion process suggest an incomplete rupture of the supraspinatus tendon. Supraspinatus, infraspinatus, teres minor and subscapularis all originate from the scapula and insert onto the fibrous capsule of the shoulder joint and either the greater or lesser tuberosity of the humerus. The tendons of insertion of these four muscles form a musculotendinous cuff about the shoulder joint called the rotator cuff. The muscles of the rotator cuff are not powerful movers of the arm. However, when powerful muscles (such as deltoid, pectoralis major, teres major and latissimus dorsi) move the arm at the shoulder joint, the muscles of the rotator cuff function to maintain the humeral head in close and proper apposition to the glenoid fossa of the scapula. In effect, the muscles of the rotator cuff serve to stabilize the dynamic integrity of the shoulder joint when the prime movers of the arm exert their forces across the joint.

    During daily abduction and flexion of the arm, the rotator cuff tendons [in particular, the supraspinatus tendon (T)] and the subacromial bursa (B) are subject to compression and friction between the head of the humerus (H) and the overlying coracoacromial ligament and acromion process (AP) (Fig. 4). The repetitive application of these disruptive forces can lead to an incomplete tear of the supraspinatus tendon, supraspinatus tendinitis, calcified depositions in the supraspinatus tendon or subacromial bursitis. Arm abduction with any of these lesions is characteristically painless from 0 to 60 degrees, painful from 60 to 120 degrees and then painless again from 120 to 180 degrees. The intermediate arc is painful because the torn part of the supraspinatus tendon or the inflamed, swollen subacromial bursa is pulled proximally and compressed beneath the acromion process and/or coracoacromial ligament during this arc of arm abduction.

Arm abduction involves movements at the shoulder, acromioclavicular and sternoclavicular joints. During the first 60 degrees abduction, much of the movement occurs within the shoulder joint. Supraspinatus and deltoid are the prime movers for about the first 15 degrees abduction from the anatomical position. Deltoid is the chief prime mover from 15 to 60 degrees abduction. Between 60 and 150 degrees abduction, much of the movement is due to lateral rotation of the scapula at the acromioclavicular joint. Trapezius and serratus anterior are the prime movers from 60 to 150 degrees abduction. Between 150 and 180 degrees abduction, much of the movement is due to lateral bending of the vertebral column.

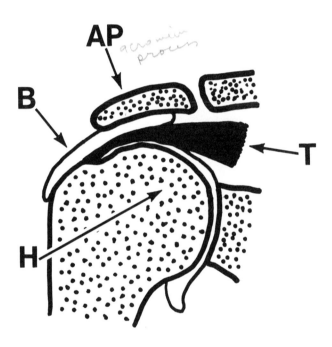

Fig. 4

A 51 year-old man suffers an injury to the right shoulder during an automobile accident. An AP radiograph of the right shoulder reveals an anterior shoulder dislocation in which the head of the humerus lies in a subglenoid position (Fig. 5).

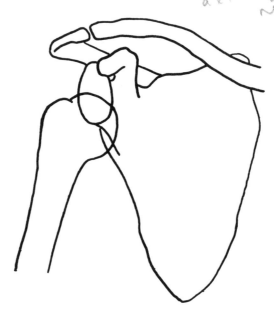

axillary nerve

Fig. 5

215. Which nerve is most likely to be injured by the inferior displacement of the humeral head?

    (A) Axillary nerve
    (B) Median nerve
    (C) Musculocutaneous nerve
    (D) Radial nerve
    (E) Ulnar nerve

216. Injury to this nerve could result in which motor or sensory deficit?

    (A) Weakness in extending the forearm at the elbow against resistance
    (B) Weakness in flexing the forearm at the elbow against resistance
    (C) Partial loss of sensation on the medial aspect of the upper arm
    (D) Partial loss of sensation on the lateral aspect of the upper arm
    (E) Weakness in pronating the forearm at the radioulnar joints against resistance

217. Which of the following arteries passes directly inferior to the shoulder joint capsule?

    (A) Thoracoacromial trunk
    (B) Anterior circumflex humeral artery
    (C) Posterior circumflex humeral artery
    (D) Lateral thoracic artery
    (E) Subscapular artery

ANSWERS AND TUTORIAL FOR ITEMS 215-217

The answers are: **215-A; 216-D; 217-C.** The axillary nerve (AN) arises from the posterior cord (PC) of the brachial plexus (Fig. 6). The axillary nerve extends toward the posterior part of the upper arm by passing posteriorly through the quadrangular space (where it accompanies the posterior circumflex humeral artery, a branch of the 3rd part of the axillary artery). The quadrangular space is a space in the upper arm bordered laterally by the surgical neck (SN) of the humerus and superiorly by the fibrous capsule of the shoulder joint (SJ). The close relation of the axillary nerve to the inferior aspect of the shoulder joint capsule renders the axillary nerve especially susceptible to injury from shoulder dislocations in which the humeral head is inferiorly displaced. The close relation of the axillary nerve to the surgical neck of the humerus renders the axillary nerve especially susceptible to injury from fractures of the surgical neck. Upon passing through the quadrangular space, the axillary nerve gives rise to the branches which innervate teres minor and deltoid. The axillary nerve also gives rise to the upper lateral cutaneous nerve of the arm. The cutaneous area (A) supplied by this nerve includes the skin overlying the lower half of deltoid (Fig. 7). The most important motor deficit that can result from injury to the axillary nerve is partial or complete loss of the actions of deltoid. Deltoid is the sole prime mover of arm abduction from 15 to 60 degrees. Injury to the axillary nerve may result in sensory deficits in the cutaneous area supplied by the upper lateral cutaneous nerve of the arm.

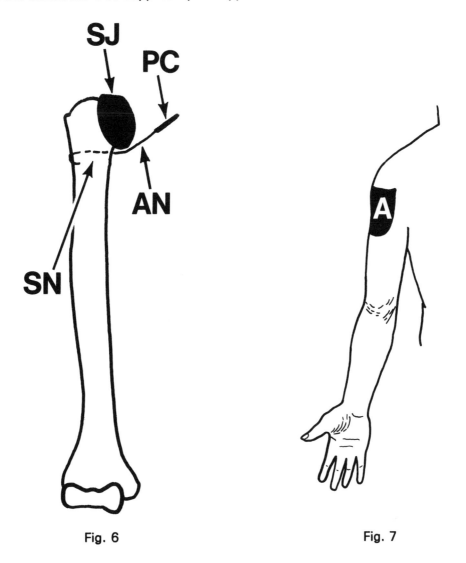

Fig. 6   Fig. 7

Items 218-220

A 28 year-old woman suffers an injury to the left arm during an automobile accident. Radiographs of the left arm show an oblique fracture through the mid-shaft of the humerus.

218. Which nerve is most likely to be injured by the mid-shaft fracture of the humerus?

(A) Axillary nerve
(B) Median nerve
(C) Musculocutaneous nerve
(D) Radial nerve
(E) Ulnar nerve

219. Injury to this nerve could result in any of the following motor deficits **EXCEPT:**

(A) weakness in extending the hand at the wrist against resistance
(B) weakness in flexing the hand at the wrist against resistance
(C) weakness in abducting the hand at the wrist against resistance
(D) weakness in adducting the hand at the wrist against resistance
(E) weakness in supinating the forearm against resistance

220. Each of the following muscles is a chief abductor or adductor of the hand **EXCEPT:**

(A) brachioradialis
(B) extensor carpi ulnaris
(C) flexor carpi radialis
(D) extensor carpi radialis longus
(E) extensor carpi radialis brevis

90

ANSWERS AND TUTORIAL FOR ITEMS 218-220

The answers are: **218-D; 219-B; 220-A.** The radial nerve (RN) arises from the posterior cord (PC) of the brachial plexus (Fig. 8). As the radial nerve extends through the mid-region of the arm (alongside the profunda brachii, a branch of the brachial artery), both the radial nerve and profunda brachii lie against the spiral groove of the shaft of the humerus. The close relation of the radial nerve to the mid-shaft of the humerus renders the radial nerve especially susceptible to injury from fractures of the mid-shaft of the humerus. Injury to the radial nerve along its course beside the spiral groove of the humerus may cause partial denervation of triceps and partial or complete denervation of any of the forearm muscles innervated by either the radial nerve (brachioradialis and extensor carpi radialis longus) or the deep branch of the radial nerve (supinator, extensor carpi radialis brevis, extensor carpi ulnaris, abductor pollicis longus, extensor pollicis brevis, extensor pollicis longus, extensor digitorum, extensor indicis and extensor digiti minimi). Weakness in extending, abducting and adducting the hand at the wrist against resistance may occur because of partial loss of action of the chief extensors of the hand (extensor carpi radialis longus, extensor carpi radialis brevis and extensor carpi ulnaris), two of the chief abductors of the hand (extensor carpi radialis longus and extensor carpi radialis brevis) and one of the chief adductors of the hand (extensor carpi ulnaris). Weakness in supinating the forearm against resistance may occur because of partial loss of action of one of the two supinators of the forearm (supinator). Weakness in flexing the hand against resistance will not occur because none of the flexors of the hand are innervated by either the radial nerve or its deep branch. Brachioradialis can flex the forearm and stabilize it in the mid-prone position. It does not exert any action across the wrist joint.

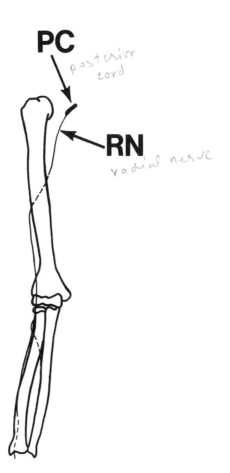

Fig. 8

91

Items 221-223

A 32 year-old woman suffers an injury to the left arm during an automobile accident. Radiographs of the left arm show a displaced fracture of the medial epicondyle of the humerus.

221. Which nerve is most likely to be injured by a displaced fracture of the medial epicondyle of the humerus?

    (A) Axillary nerve
    (B) Median nerve
    (C) Musculocutaneous nerve
    (D) Radial nerve
    (E) Ulnar nerve

222. Injury to this nerve could result in any of the following motor deficits **EXCEPT**:

    (A) weakness in holding a piece of paper between the ring and little fingers against resistance
    (B) weakness in holding a piece of paper between the middle and ring fingers against resistance
    (C) weakness in holding a piece of paper between the index and middle fingers against resistance
    (D) weakness in flexing the distal phalanx of the little finger against resistance
    (E) weakness in flexing the distal phalanx of the thumb against resistance

223. Which muscle can adduct the ring finger?

    (A) 2nd palmar interosseous
    (B) 3rd palmar interosseous
    (C) 1st dorsal interosseous
    (D) 2nd dorsal interosseous
    (E) 3rd dorsal interosseous

ANSWERS AND TUTORIAL FOR ITEMS 221-223

The answers are: **221-E; 222-E; 223-B.** The ulnar nerve (UN) arises from the medial cord (MC) of the brachial plexus (Fig. 9). As the ulnar nerve extends across the elbow region, it passes directly posterior to the medial epicondyle (ME) of the humerus. The close relation of the ulnar nerve to the medial epicondyle of the humerus renders the ulnar nerve especially susceptible to injury from displaced fractures of the medial epicondyle.

Injury to the ulnar nerve from a displaced fracture of the medial epicondyle of the humerus may cause partial or complete denervation of the forearm muscles innervated by the ulnar nerve (flexor carpi ulnaris and the medial half of flexor digitorum profundus) and/or any of the intrinsic hand muscles innervated by the deep branch of the ulnar nerve (the palmar and dorsal interossei, the 3rd and 4th lumbricals, abductor digiti minimi, flexor digiti minimi, opponens digiti minimi and adductor pollicis). Weakness in holding a piece of paper between any two fingers may occur because of partial loss of action of one or more of the abductors and adductors of the fingers (the palmar and dorsal interossei and abductor digiti minimi). Weakness in flexing the distal phalanx of the little finger against resistance may occur because of partial loss of action of the only flexor of the distal phalanx of the little finger (the muscular part of flexor digitorum profundus that pulls upon the tendon that inserts onto the base of the distal phalanx of the little finger). Weakness in flexing the distal phalanx of the thumb against resistance will not occur because the sole flexor of the distal phalanx of the thumb (flexor pollicis longus) is innervated by the anterior interosseous branch of the median nerve.

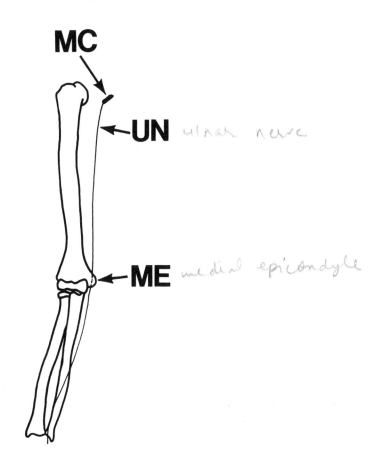

Fig. 9

carpal tunnel syndrome (handwritten annotation)

Items 224-226

A 51 year-old woman complains of recent, progressive clumsiness in sewing and writing with her right hand and reports numbness and tingling in her right hand upon awakening in the morning from sleep. She says that shaking the right hand relieves the numbness and tingling. Examination reveals slight atrophy of the muscle mass underlying the thenar eminence of the right hand and difficulty in drawing the thumb of the right hand across the palm in order to touch the base of the little finger with the tip of the thumb. Light percussion on the lateral (radial) side of the palmaris longus tendon produces a tingling sensation in the right hand whose distribution is similar to that experienced by the patient upon awakening in the morning.

224. Which nerve is impaired in the patient's right upper limb?

   (A) Anterior interosseous branch of the median nerve
   (B) Median nerve
   (C) Ulnar nerve
   (D) Deep branch of the ulnar nerve
   (E) Superficial branch of the ulnar nerve

225. The patient's condition may produce sensory deficits in any of the following cutaneous areas **EXCEPT:**

   (A) palmar (volar) surface of the little finger
   (B) palmar surface of the middle finger
   (C) palmar surface of the index finger
   (D) palmar surface of the thumb
   (E) dorsal surface of the distal third of the index finger

226. All of the following structures cross the wrist by extending anterior to the flexor retinaculum **EXCEPT:**

   (A) palmar cutaneous branch of the median nerve
   (B) palmaris longus tendon
   (C) flexor pollicis longus tendon
   (D) ulnar nerve
   (E) ulnar artery

ANSWERS AND TUTORIAL FOR ITEMS 224-226

The answers are: **224-B; 225-A; 226-C.** The signs and symptoms of the patient's motor and sensory deficits in the right hand are characteristic of carpal tunnel syndrome. The carpal tunnel syndrome is produced by conditions which compress the median nerve as it passes through the carpal tunnel. The carpal tunnel is an osseofascial corridor in the wrist bordered anteriorly by the flexor retinaculum (FR) and posteriorly by the carpal bones (CB) (Fig. 10). The median nerve (MN), the tendons of flexor digitorum superficialis and profundus and the tendon of flexor pollicis longus (FPLT) all cross the wrist through the carpal tunnel. In contrast, the ulnar nerve (UN), ulnar artery (UA), palmaris longus tendon (PLT) and the palmar cutaneous branch of the median nerve (PC) all cross the wrist anterior to the flexor retinaculum. The median nerve is susceptible to compression neuropathy by conditions which increase the pressure on the contents of the carpal tunnel.

94

The median nerve innervates five intrinsic muscles of the hand: the muscles of the thenar eminence (abductor pollicis brevis, flexor pollicis brevis and opponens pollicis) and the 1st and 2nd lumbricals. The median nerve also supplies cutaneous areas of the hand. The blocked areas of Figs. 11A and 11B respectively show the areas on the volar and dorsal surfaces of the hand innervated by cutaneous branches of the median nerve. However, it is important to note that the palmar cutaneous branch of the median nerve, which is the cutaneous nerve that provides sensory innervation for the radial two-thirds of the palmar surface of the hand, arises from the median nerve proximal to the carpal tunnel and crosses the wrist anterior to the flexor retinaculum. Severe compression neuropathy of the median nerve in the carpal tunnel can thus lead to atrophy of the muscles underlying the thenar eminence and weakness of thumb opposition. Opponens pollicis is the prime mover of thumb opposition. The little finger is the sole digit whose cutaneous innervation is completely spared in individuals suffering from carpal tunnel syndrome.

The digital pain and/or paresthesia associated with the carpal tunnel syndrome can be reproduced during physical examination by tapping a reflex hammer on the radial side of the palmaris longus tendon along the tendon's course over the flexor retinaculum (Tinel's sign). The median nerve (MN) in the carpal tunnel lies deep to the radial margin of the palmaris longus tendon (PLT) (Fig. 10). In individuals suffering from carpal tunnel syndrome, the light percussion momentarily increases the pressure upon an already compressed median nerve and thus may elicit pain and/or paresthesia in the areas of the lateral three and a half digits supplied by the digital cutaneous branches of the median nerve.

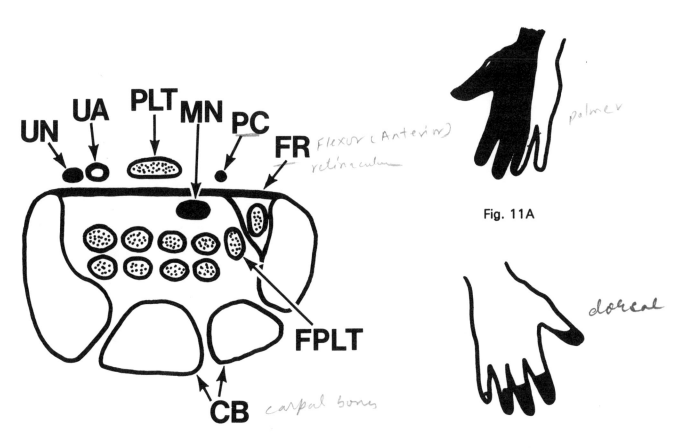

Fig. 11A

Fig. 11B

Fig. 10

A 24 year-old man is thrown out of a pick-up truck during an accident and sustains an injury to the left upper limb. Examination at the emergency center does not reveal any cuts, fractures, or dislocations in the left upper limb. However, a number of motor and sensory deficits are found which suggest that the patient has sustained a traction injury to one or more parts of the brachial plexus.

227. Any of the following motor deficits would be consistent with a traction injury to the upper trunk of the brachial plexus EXCEPT:

   (A) weakness in bracing the shoulder
   (B) weakness in rotating the arm laterally against resistance
   (C) weakness in initiating arm abduction from the anatomical position
   (D) weakness in flexing the arm against resistance
   (E) weakness in supinating the forearm against resistance

228. Which labeled cutaneous area in the illustration below is not a part of the dermatomes of the cutaneous sensory fibers that pass through the upper trunk of the brachial plexus?

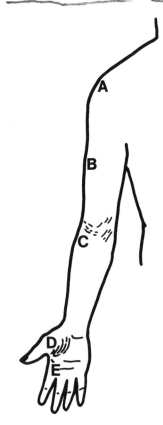

229. Any of the following motor deficits would be consistent with a traction injury to the lower trunk of the brachial plexus **EXCEPT**:

    (A) weakness in flexing the distal phalanx of the little finger against resistance
    (B) weakness in abducting the little finger
    (C) weakness in flexing the distal phalanx of the thumb against resistance
    (D) weakness in abducting the thumb
    (E) weakness in supinating the forearm against resistance

230. The biceps brachii stretch reflex involves sensory and motor nerve fibers in primarily spinal nerves

    (A) C5 and C6
    (B) C6 and C7
    (C) C7 and C8
    (D) C8 and T1
    (E) T1 and T2

231. The triceps brachii stretch reflex involves sensory and motor nerve fibers in primarily spinal nerves

    (A) C5 and C6
    (B) C6 and C7
    (C) C7 and C8
    (D) C8 and T1
    (E) T1 and T2

232. The cell bodies of the sensory fibers that provide cutaneous innervation for the medial aspect of the arm are located in the dorsal root ganglia of spinal nerves

    (A) C5 and C6
    (B) C6 and C7
    (C) C7 and C8
    (D) C8 and T1
    (E) T1 and T2

ANSWERS AND TUTORIAL FOR ITEMS 227-232

The answers are: **227-A; 228-E; 229-E; 230-A; 231-C; 232-D.** The upper trunk of the brachial plexus is formed from the union of the C5 and C6 roots. A traction injury to the upper trunk chiefly affects those upper limb movements whose prime movers are muscles innervated exclusively by C5 and C6 fibers (supraspinatus, infraspinatus, teres minor and deltoid) or predominantly by C5 and C6 fibers (biceps brachii, brachialis, brachioradialis and supinator). Deltoid, infraspinatus and teres minor are the sole lateral rotators of the arm. Supraspinatus and deltoid initiate arm abduction from the anatomical position. Biceps brachii and brachialis are the chief flexors of the forearm. Biceps brachii and supinator are the sole supinators of the forearm. Trapezius is the chief retractor of the scapula and thus the prime mover for bracing the shoulder. Levator scapulae, rhomboid major and rhomboid minor assist trapezius in this action. Trapezius is innervated by the spinal part of the accessory nerve. Levator scapulae is innervated by C3 and C4 fibers and the dorsal scapular nerve. The rhomboids are innervated by the dorsal scapular nerve. The dorsal scapular nerve arises from the C5 root of the brachial plexus.

The lower trunk of the brachial plexus is formed from the union of the C8 and T1 roots. A traction injury to the lower trunk chiefly affects those upper limb movements whose prime movers are muscles innervated either exclusively or predominantly by C8 and/or T1 fibers (flexor carpi ulnaris, flexor digitorum superficialis, flexor digitorum profundus, flexor pollicis longus, pronator quadratus, extensor carpi ulnaris, abductor pollicis longus, extensor pollicis brevis, extensor pollicis longus, extensor indicis and all of the intrinsic muscles of the hand). A traction injury to the lower trunk of the brachial plexus can affect all movements of the hand and its digits. Flexor digitorum profundus is the sole flexor of the distal phalanx of the little finger. Abductor digiti minimi is the sole abductor of the little finger. Flexor pollicis longus is the sole flexor of the distal phalanx of the thumb. Abductor pollicis longus and abductor pollicis brevis are the sole abductors of the thumb.

Figs. 12A and 12B show, respectively, the dermatomes on the anterior and posterior surfaces of the upper limb.

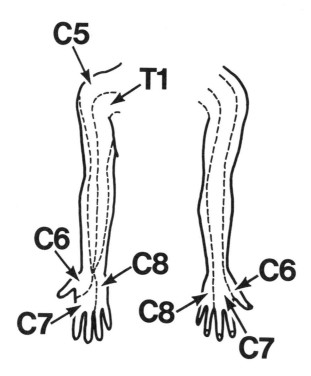

Fig. 12A   Fig. 12B

Items 233-238

An 8 year-old girl's right hand is clawed by a cat. Examination of the child a day later reveals evidence of lymphatic dissemination of a hand infection.

233. If superficial tissues on the medial aspect of the palm of the girl's hand are infected, which of the following groups of lymph nodes will commonly be the first to react to lymphatic dissemination of the infection?

    (A) Anterior group of axillary nodes
    (B) Lateral group of axillary nodes
    (C) Deltopectoral (infraclavicular) group of axillary nodes
    (D) Central group of axillary nodes
    (E) Supratrochlear nodes

234. If deep tissues of the hand are infected, which of the following groups of lymph nodes will commonly be the first to react to lymphatic dissemination of the infection?

    (A) Anterior group of axillary nodes
    (B) Lateral group of axillary nodes
    (C) Deltopectoral (infraclavicular) group of axillary nodes
    (D) Central group of axillary nodes
    (E) Supratrochlear nodes

235. Superficial lymph nodes which are mounting an immunological response to infectious agents generally exhibit all of the following characteristics **EXCEPT:**

    (A) The nodes are enlarged.
    (B) The nodes are irregularly shaped.
    (C) The nodes are tender.
    (D) The nodes are relatively mobile.
    (E) The nodes have a firm consistency.

236. Which group of axillary lymph nodes is clustered about the region where the cephalic vein pierces deep fascia to extend toward its union with the axillary vein?

    (A) Anterior group of axillary nodes
    (B) Posterior group of axillary nodes
    (C) Lateral group of axillary nodes
    (D) Central group of axillary nodes
    (E) Deltopectoral (infraclavicular) group of axillary nodes

237. Which group of axillary lymph nodes is palpable against the anterior aspect of the posterior axillary fold?

    (A) Anterior group of axillary nodes
    (B) Posterior group of axillary nodes
    (C) Lateral group of axillary nodes
    (D) Central group of axillary nodes
    (E) Deltopectoral (infraclavicular) group of axillary nodes

238. Which group of axillary lymph nodes is the first to receive lymph drained from the lateral half of the mammary gland?

    (A) Anterior group of axillary nodes
    (B) Posterior group of axillary nodes
    (C) Lateral group of axillary nodes
    (D) Central group of axillary nodes
    (E) Deltopectoral (infraclavicular) group of axillary nodes

ANSWERS AND TUTORIAL FOR ITEMS 233-238

The answers are: **233-E; 234-B; 235-B; 236-E; 237-B; 238-A.** The supratrochlear lymph nodes can be palpated along the medial side of the basilic vein immediately superior to the medial epicondyle of the humerus. The supratrochlear nodes drain superficial tissues of the medial aspect of the hand and forearm. The lateral group of axillary lymph nodes can be palpated along the humeral head. The lateral group of axillary nodes drains all of the deep tissues of the hand, forearm and arm.

Lymph nodes which are mounting an immunological response to infectious agents are enlarged but regularly shaped, tender, relatively mobile and firm. Normal superficial lymph nodes are frequently difficult to palpate because of their small size and are non-tender, relatively mobile and soft.

The deltopectoral group of axillary nodes drains superficial tissues on the lateral aspect of the hand, forearm and arm. The posterior (subscapular) group of axillary nodes drains superficial tissues of the posterior aspect of the trunk, down to the level of the iliac crest. The anterior (pectoral) group of axillary nodes can be palpated against the posterior aspect of the anterior axillary fold.

In the following items, match each carpal bone with its image in the labeled PA radiograph of the hand.

239. Scaphoid

240. Capitate

241. Lunate

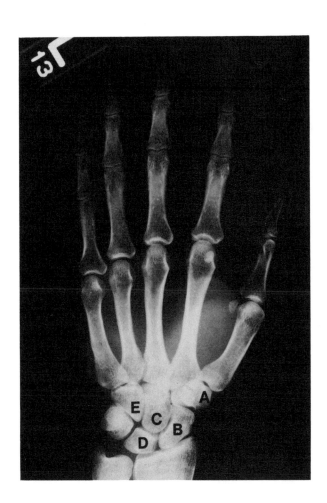

ANSWERS AND TUTORIAL FOR ITEMS 239-241

The answers are: **239-B; 240-C; 241-D.** The carpal bone labeled A is the trapezium and the carpal bone labeled E is the hamate. The trapezium, scaphoid and styloid process of the radius form the bony floor of the anatomical snuffbox of the wrist. The capitate is the largest carpal bone. Note that the central axes of the 3rd metacarpal and capitate are co-linear. In the wrist joint, the scaphoid, lunate and triquetrum articulate with the distal end of the radius and the triangular disk of fibrous cartilage.

The following items address information an examiner needs to know when examining the hip joint.

242. All of the following statements concerning the limits of passive movement about the hip joint of a normal young adult are correct EXCEPT:

(A) The thigh can be externally rotated approximately 45 degrees.
(B) The thigh can be internally rotated approximately 40 degrees.
(C) If the leg is flexed 90 degrees at the knee, the thigh can be passively flexed approximately 120 degrees.
(D) If the leg is fully extended (is at 0 degrees flexion at the knee), the thigh can be passively flexed approximately 45 degrees.
(E) The thigh can be abducted approximately 45 degrees.

243. The most powerful flexor of the thigh is

(A) sartorius
(B) pectineus
(C) iliopsoas
(D) rectus femoris
(E) adductor longus

244. A chief abductor of the thigh is

(A) piriformis
(B) gracilis
(C) quadratus femoris
(D) gluteus medius & minimus
(E) gluteus maximus

245. All of the following muscles are extensors of the thigh EXCEPT:

(A) obturator externus
(B) hamstring part of adductor magnus
(C) semitendinosus
(D) long head of biceps femoris
(E) semimembranosus

246. All of the following statements concerning the innervation of muscles that move the thigh are correct EXCEPT:

(A) The most powerful flexor of the thigh is innervated by nerves derived from the lumbar plexus.
(B) All of the extensors of the thigh are innervated by nerves derived from the sacral plexus.
(C) The chief abductors of the thigh are innervated by nerves derived from the lumbar plexus.
(D) All of the adductors of the thigh are innervated by nerves derived from the lumbar plexus.
(E) All of the medial rotators of the thigh are innervated by nerves derived from the sacral plexus.

247. All of the following disorders can produce a positive Trendelenburg sign EXCEPT:

(A) paralysis of gluteus medius and gluteus minimus
(B) an abnormal angle of inclination between the neck and shaft of the femur
(C) dislocation of the hip joint
(D) ischial bursitis
(E) osteoarthritis of the hip joint

ANSWERS AND TUTORIAL FOR ITEMS 242-247

The answers are: **242-D; 243-C; 244-D; 245-A; 246-C; 247-D.** If the leg is fully extended at the knee, the thigh can be passively flexed approximately 90 degrees before tension in the hamstring muscles becomes painful and limits further flexion. Iliopsoas is the most powerful flexor of the thigh. Iliacus is innervated by the femoral nerve and psoas major is innervated by nerve fibers from L1, L2 and L3.

The extensors of the thigh and their innervation are as follows:
Gluteus maximus - inferior gluteal nerve
Semitendinosus - tibial portion of the sciatic nerve
Semimembranosus - tibial portion of the sciatic nerve
Long head of biceps femoris - tibial portion of the sciatic nerve
Hamstring part of adductor magnus - tibial portion of the sciatic nerve

The chief abductors of the thigh, gluteus medius and gluteus minimus, are innervated by the superior gluteal nerve.

The adductors of the thigh and their innervation are as follows:
Adductor part of adductor magnus - obturator nerve
Adductor longus - obturator nerve
Adductor brevis - obturator nerve
Pectineus - obturator and femoral nerves
Gracilis - obturator nerve

The medial rotators of the thigh, tensor fasciae latae, gluteus minimus and gluteus medius, are innervated by the superior gluteal nerve.

When an individual attempts to stand (or support the body) on one lower limb only, gluteus medius and minimus (the chief abductors of the thigh) on the side of the supporting lower limb tense to elevate the contralateral side of the pelvis. This tilting of the pelvis balances the body over the supporting lower limb. An inability to tilt the pelvis in this fashion is called a positive Trendelenburg sign. Any disorder which limits the capacity of gluteus medius and gluteus minimus to support and steady the body when standing on the ipsilateral lower limb or produces pain in the hip joint when it bears the weight of the upper body may produce a positive Trendelenburg sign. Ischial bursitis is inflammation of the ischial bursa. The ischial bursa lies directly superficial to the ischial tuberosity.

Items 248-252

The parents of a 6 year-old boy report that their son has started walking with a limp. The pediatrician learns that the boy has been in good health for the past few months. His sublingual temperature is 98.5° F. The patient states that he walks with a limp because of right hip pain that started a week ago. The hip pain is aggravated by weight-bearing and relieved by rest. Trendelenburg's sign is observed when the boy is asked to stand on the right lower limb only. Radiographs of the hip joints show periarticular soft tissue swelling about the right hip joint. A bone scan reveals diminished activity in the anterolateral aspect of the right capital femoral epiphysis.

248. The history, physical exam and radiological findings suggest that the patient has

    (A) transient synovitis of the hip joint
    (B) septic arthritis of the hip joint
    (C) Legg-Calve-Perthes disease
    (D) a slipped capital femoral epiphysis
    (E) epiphyseal dysplasia

249. The patient's pain may be referred to any of the following regions EXCEPT:

    (A) groin
    (B) anterior aspect of the thigh
    (C) medial aspect of the thigh
    (D) medial aspect of the knee
    (E) leg

250. The arteries which are the chief source of blood supply to the patient's capital femoral epiphysis are

    (A) arteries in the ligament to the head of the femur
    (B) medullary arteries of the metaphysis
    (C) arteries derived from the medial and lateral circumflex femoral arteries
    (D) arteries derived from the superior gluteal artery
    (E) arteries derived from the inferior gluteal artery

251. Which of the following positions will the patient prefer to assume in order to minimize the lower limb pain?

    (A) The right thigh completely extended and slightly externally rotated
    (B) The right thigh completely extended and neutrally rotated
    (C) The right thigh completely extended and slightly internally rotated
    (D) The right thigh flexed 90 degrees and slightly externally rotated
    (E) The right thigh flexed 90 degrees and slightly internally rotated

252. All of the following nerves innervate the hip joint EXCEPT:

    (A) superior gluteal nerve
    (B) pudendal nerve
    (C) femoral nerve
    (D) obturator nerve
    (E) sciatic nerve

ANSWERS AND TUTORIAL FOR ITEMS 248-252

The answers are: 248-C; 249-E; 250-C; 251-D; 252-B. Legg-Calve-Perthes disease is a self-limiting hip disorder of children involving vascular compromise of the capital femoral epiphysis. Diminished vascularization in the anterolateral aspect of the capital femoral epiphysis is characteristic of the pattern of ischemic necrosis which occurs in Legg-Calve-Perthes disease. The radiographic evidence of periarticular soft tissue swelling suggests that the patient is also suffering from inflammation of the synovial membrane and effusion of the hip joint.

The etiology of Legg-Calve-Perthes disease is unknown. The anatomical basis of the disorder is believed to be related to the developmental changes in the blood supply to the hip during childhood. Branches of the medial and lateral circumflex femoral arteries form an extracapsular vascular ring around the base of the neck of the femur. Branches of the extracapsular vascular ring called the retinacular arteries ascend along the neck of the femur to penetrate the capsule of the hip joint and give rise within the joint to branches that supply the upper end of the femur. The retinacular arteries and their branches are believed to be the chief source of blood supply to the head of the femur at all ages. At birth, the cartilaginous femoral head is also supplied by arteries entering the head from the shaft of the femur. Establishment of an epiphyseal plate between the capital epiphysis and the metaphysis by about 4 years of age abolishes all blood supply from the metaphysis. This arrangement persists until about 9 years of age, at which time the arteries in the ligament to the head of the femur begin to become increasingly prominent in supplying the head of the femur. Medullary arteries of the metaphysis begin to supply the femoral head upon epiphyseal fusion in late adolescence or early adulthood. The childhood period from 4 to 9 years of age, when the retinacular arteries are the sole significant source of blood supply to the capital epiphysis, spans the years of the highest incidence of Legg-Calve-Perthes disease.

Hip joint disease frequently refers pain to the groin, anteromedial aspects of the thigh and the medial aspect of the knee. Cutaneous branches of the obturator nerve supply the medial aspect of the thigh. Cutaneous branches of the femoral nerve supply the anterior aspect of the thigh and the anteromedial aspect of the knee. A major component of the patient's right hip pain is emanating from tension on the joint's inflamed synovial membrane. Synovial membrane tension in the hip joint is minimized when the thigh is flexed 90 degrees and slightly externally rotated. Accordingly, patients with a painful hip joint effusion are generally most comfortable when seated with the painful thigh slightly externally rotated. Active and passive attempts at internal rotation or abduction of the thigh commonly exacerbate the lower limb pain of patients with Legg-Calve-Perthes disease.

Items 253-256

The following items address information an examiner needs to know when examining the knee joint.

253. All of the following statements concerning the limits of passive movement about the knee joint of a normal young adult are correct EXCEPT:

    (A) The leg can be flexed approximately 130 degrees.
    (B) The leg can be hyperextended 10-15 degrees.
    (C) If the leg is flexed 90 degrees, it can be internally rotated approximately 30 degrees.
    (D) If the leg is flexed 90 degrees, it can be externally rotated approximately 40 degrees.
    (E) If the leg is fully extended, it can be internally or externally rotated approximately 45 degrees.

254. The quadriceps femoris stretch reflex involves sensory and motor nerve fibers in spinal nerves

    (A) L1, L2 and L3
    (B) L2, L3 and L4
    (C) L3, L4 and L5
    (D) L4, L5 and S1
    (E) L5, S1 and S2

255. All of the following nerves innervate one or more flexors of the leg EXCEPT:

    (A) deep peroneal nerve
    (B) common peroneal portion of the sciatic nerve
    (C) tibial portion of the sciatic nerve
    (D) femoral nerve
    (E) obturator nerve

256. An effusion of the knee joint's synovial cavity may result in all of the following EXCEPT:

    (A) a swelling of the soft tissues directly anterior to the patella
    (B) a swelling along the upper margin of the patella
    (C) swellings along the medial or lateral sides of the patella
    (D) swellings along the medial or lateral sides of the ligamentum patellae
    (E) displacement of the patella anteriorly from the intercondylar area of the femur

ANSWERS AND TUTORIAL FOR ITEMS 253-256

The answers are: **253-E; 254-B; 255-A; 256-A.** The knee joint is in its most stable configuration when fully extended (at 0 degrees flexion). When the knee joint is fully extended, the femoral condyles are in maximum congruence and contact with the meniscotibial surfaces. There is maximum compression between the femoral and meniscotibial surfaces because the anterior and posterior cruciate ligaments, the medial and lateral collateral ligaments and the oblique popliteal ligament are all in part twisted and tightly stretched across the knee. The leg is externally rotated to the maximum extent within the knee joint and thus no internal or external rotation of the leg is possible. Assessment of passive internal and external rotation of the leg at the knee is conducted with the leg flexed 90 degrees.

The muscles of quadriceps femoris (rectus femoris, vastus medialis, vastus intermedius and vastus lateralis) are the sole extensors of the leg. They are all innervated by the femoral nerve. The femoral nerve is derived from the L2, L3 and L4 roots of the lumbar plexus.

The flexors of the leg and their innervation are as follows:
 Semitendinosus - tibial portion of the sciatic nerve
 Semimembranosus - tibial portion of the sciatic nerve
 Long head of biceps femoris - tibial portion of the sciatic nerve
 Short head of biceps femoris - common peroneal portion of the sciatic nerve
 Gastrocnemius - tibial nerve
 Popliteus - tibial nerve
 Plantaris - tibial nerve

When fluid accumulates in the knee joint's synovial cavity, the normal hollowed contours around the patella and along the sides of the ligamentum patellae may swell and the patella may be displaced anteriorly. The contours surrounding the upper margin of the patella may swell because of expansion of the underlying suprapatellar bursa. The suprapatellar bursa is an extension of the knee joint's synovial cavity. The bursa lies deep to the quadriceps femoris tendon and extends for about 5-7 cm above the upper border of the patella. The hollowed contours along the sides at the ligamentum patellae may swell because the knee joint's synovial membrane lining extends inferiorly along both sides of the infrapatellar fat pad, which lies deep to the ligamentum patellae. The patella may be displaced anteriorly because the patella and the portion of the quadriceps femoris tendon in which it is embedded form the anterior aspect of the knee joint. The subcutaneous tissues that overlie the patella lie superficial to the knee joint and thus do not swell when the knee joint has an effusion. The prepatellar bursa lies in the subcutaneous tissues that overlie the patella, but the prepatellar bursa does not communicate with the knee joint's synovial cavity.

Items 257-259

A 20 year-old man receives a severe blow to the lateral side of the right knee while playing football. Examination two hours later reveals a stiffly swollen knee. A variety of orthopedic tests are conducted upon aspiration of the hemorrhagic effusion of the right knee joint. Valgus stress applied at the knee with the knee at 30 degrees flexion produces valgus subluxation of the tibia on the femur. Valgus stress applied at the knee with the fully extended knee also produces valgus subluxation of the tibia on the femur. Varus stress applied at the knee with the knee at 30 degrees flexion does not produce a varus subluxation of the tibia on the femur. The anterior drawer test with the knee at 90 degrees flexion produces a 3 cm anterior subluxation of the tibia on the femur. The posterior drawer test with the knee at 90 degrees flexion produces about 0.5 cm posterior subluxation of the tibia on the femur. McMurray's test with the leg externally rotated produces a painful, audible click upon leg extension.

257. All of the following structures are significantly torn in the patient's right knee **EXCEPT**:

    (A) anterior cruciate ligament
    (B) posterior cruciate ligament
    (C) medial collateral ligament
    (D) medial meniscus
    (E) fibrous capsule of the knee joint

258. All of the following statements regarding the ligaments and cartilages of the knee joint are correct **EXCEPT**:

    (A) The lateral collateral ligament extends from the lateral epicondyle of the femur to the lateral condyle of the tibia.
    (B) The medial collateral ligament extends from the medial epicondyle of the femur to the medial condyle and upper part of the shaft of the tibia.
    (C) The anterior cruciate ligament extends from the anterior area of the tibial intercondylar surface to the medial surface of the lateral femoral condyle.
    (D) The posterior cruciate ligament extends from the posterior area of the tibial intercondylar surface to the lateral surface of the medial femoral condyle.
    (E) The horns of the menisci are attached to the tibial condyle below.

259. The ligament that restricts the tibia from being pulled too far anteriorly during extension of the leg is the

    (A) medial collateral ligament
    (B) lateral collateral ligament
    (C) anterior cruciate ligament
    (D) posterior cruciate ligament
    (E) oblique popliteal ligament

ANSWERS AND TUTORIAL FOR ITEMS 257-259

The answers are: **257-B; 258-A; 259-C.** The direction of the blow to the right knee indicates that the structures which provide medial stability for the knee have sustained damage. The rapid onset of an hemorrhagic effusion indicates second or third degree sprains of the joint's capsular and ligamentous supports.

The valgus and varus stress tests assess the integrity of the structures which provide, respectively, medial and lateral stability for the knee. As used here, the term valgus indicates a laterally-directed stress applied to the lower end of the leg in conjunction with a medially-directed stress applied at the knee, and the opposite term varus indicates a medially-directed stress applied to the lower end of the leg in conjunction with a laterally-directed stress applied at the knee. When the knee is at 30 degrees flexion, the anterior and posterior cruciate ligaments are lax and the medial and lateral collateral ligaments are the most effective ligamentous supports, respectively, of medial and lateral stability of the knee joint. When the knee is fully extended, the anterior and posterior cruciate ligaments are taut and can by themselves provide medial and lateral stability in the absence of collateral ligament support.

The negative finding obtained upon applying a varus stress with the knee at 30 degrees flexion indicates the absence of any significant damage to the lateral collateral ligament. The subluxation obtained upon applying a valgus stress with the knee at 30 degrees flexion suggests significant damage to the medial collateral ligament. The subluxation obtained upon applying a valgus stress with the knee at 0 degrees flexion suggests significant damage to at least one of the following structures: anterior cruciate ligament, posterior cruciate ligament, or posteromedial aspect of the fibrous capsule.

The anterior and posterior drawer tests assess the integrity of the cruciate ligaments and the fibrous capsule. Test results of 1 cm or less of anterior or posterior subluxation of the tibia on the femur are considered to be negative findings. However, test results on the injured knee should always be compared with those on the contralateral, uninjured knee. The positive anterior drawer test suggests significant damage to the anterior cruciate ligament and one or more parts of the fibrous capsule. The negative posterior drawer test suggests that there is no significant damage to the posterior cruciate ligament. The findings of the anterior and posterior drawer tests thus indicate that the subluxation obtained upon applying a valgus stress with the knee at 0 degrees flexion is due to third degree sprains of the anterior cruciate ligament and fibrous capsule.

McMurray's test is used to detect tears of the menisci. As the patient lie supine with the thigh and leg flexed, the examiner grasps and presses on the medial and lateral aspects of the knee with one hand and holds the plantar surface of the foot with the other hand. The examiner flexes the leg until the heel almost touches the buttock. The examiner externally rotates the leg if the patient is suspected of having a tear of the medial meniscus (as in this case) or internally rotates the leg if the patient is suspected of having a tear of the lateral meniscus. External rotation of the leg draws the medial meniscus toward the center of the joint. Internal leg rotation draws the lateral meniscus toward the center of the joint. While forcefully maintaining the leg in either rotated configuration, the examiner extends the leg to 90 degrees flexion at the knee joint. The test is positive for a meniscal tear if the leg extension maneuver produces a palpable or audible click in association with pain. It is believed that the click occurs when the femoral condyle rolls over the meniscal tear. McMurray's test may reveal tears in the posterior and middle thirds at the meniscus but not in the anterior third.

Ankle & Foot Jt.

Items 260-264

The following items address information an examiner needs to know when examining the ankle and foot.

In items 260-262, match each movement of the foot with the pair of muscles chiefly responsible for the movement.

    (A) Tibialis anterior and tibialis posterior
    (B) Plantaris and quadratus plantae
    (C) Gastrocnemius and soleus
    (D) Peroneus longus and peroneus brevis
    (E) Extensor digitorum longus and flexor digitorum longus

260. Plantarflexion of the foot

261. Inversion of the foot

262. Eversion of the foot

263. The pulsations of the posterior tibial artery can be palpated

    (A) on the dorsum of the foot immediately lateral to the tendon of extensor hallucis longus
    (B) posteroinferiorly to the lateral malleolus
    (C) anterosuperiorly to the lateral malleolus
    (D) posteroinferiorly to the medial malleolus
    (E) anterosuperiorly to the medial malleolus

264. The Achilles tendon stretch reflex involves sensory and motor nerve fibers in spinal nerves

    (A) L1, L2 and L3
    (B) L2, L3 and L4
    (C) L3, L4 and L5
    (D) L4, L5 and S1
    (E) L5, S1 and S2

ANSWERS AND TUTORIAL FOR ITEMS 260-264

The answers are: **260-C; 261-A; 262-D; 263-D; 264-E.** Gastrocnemius and soleus are the chief plantarflexors of the foot. Tibialis posterior, flexor digitorum longus, flexor hallucis longus, peroneus longus and peroneus brevis all assist in this action. Tibialis anterior and tibialis posterior are the chief invertors of the foot. Extensor hallucis longus assists in this action. Peroneus longus and peroneus brevis are the chief evertors of the foot. Peroneus tertius assists in this action. The chief muscles which insert onto the calcaneus via the Achilles tendon (gastrocnemius and soleus) are innervated by fibers of the tibial nerve derived from the L5, S1 and S2 roots of the sacral plexus.

A 29 year-old man sustains a severe compression injury to his left upper leg ás a result of a fall from a motorcycle. Examination reveals a contusion surrounding the head and neck of the fibula.

265. Which nerve is susceptible to direct injury by a severe compression force applied to the lateral aspect of the head and neck of the fibula?

   (A) Common peroneal nerve
   (B) Deep peroneal nerve
   (C) Superficial peroneal nerve
   (D) Tibial nerve
   (E) Saphenous nerve

266. Injury to this nerve could result in weakness in all of the following movements EXCEPT:

   (A) inversion of the foot
   (B) eversion of the foot
   (C) dorsiflexion of the foot
   (D) flexion of the toes
   (E) extension of the toes

267. Cutaneous branches of the deep peroneal nerve supply the

   (A) medial side of the big toe
   (B) adjacent sides of the big and 2nd toes
   (C) adjacent sides of the 2nd and 3rd toes
   (D) adjacent sides of the 3rd and 4th toes
   (E) adjacent sides of the 4th and 5th toes

268. Which nerve's cutaneous branches supply the dorsum of the foot?

   (A) Superficial peroneal nerve
   (B) Deep peroneal nerve
   (C) Tibial nerve
   (D) Sural nerve
   (E) Saphenous nerve

ANSWERS AND TUTORIAL FOR ITEMS 265-268

The answers are: **265-A; 266-D; 267-B; 268-A.** As the common peroneal nerve descends through the popliteal fossa, it parallels the medial border of biceps femoris's tendon of insertion. The common peroneal nerve (CPN) then curves inferolaterally around the head (H) and neck (N) of the fibula as it enters the leg (Fig. 13). The common peroneal nerve ends in this vicinity by dividing into the superficial and deep peroneal nerves. The superficial peroneal nerve innervates peroneus longus and peroneus brevis, the major evertors of the foot. The deep peroneal nerve innervates all the dorsiflexors of the foot (tibialis anterior, extensor hallucis longus, extensor digitorum longus and peroneus tertius) and extensor digitorum brevis. Tibialis anterior is one of the two major invertors of the foot. Extensor hallucis longus, extensor digitorum longus and extensor digitorum brevis are the extensors of the toes. The superficial and deep peroneal nerves do not innervate any of the flexors of the toes. The flexors of the toes are all innervated by the tibial nerve or one its two terminal branches, the medial and lateral plantar nerves. The adjacent sides of the big and 2nd toes are the only cutaneous areas of the foot supplied by cutaneous branches of the deep peroneal nerve. Cutaneous branches of the superficial peroneal nerve supply the dorsum of the foot. Cutaneous branches of the medial and lateral plantar nerves supply most of the sole of the foot.

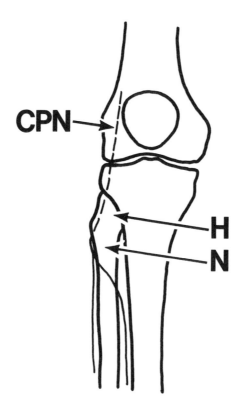

Fig. 13

A 61 year-old man states that he likes to walk for exercise, but that he recently has pain in his right calf that almost always appears after 10 minutes of walking. Resting for a few minutes alleviates the pain.

269. The most likely diagnosis for the patient's calf pain is

    (A) an arterial aneurysm in the right lower limb
    (B) deep venous thrombosis of the right lower limb
    (C) incompetent valves in the deep veins of the right leg
    (D) atherosclerotic occlusive disease of the right lower limb
    (E) a tear of one of the heads of gastrocnemius in the right lower limb

270. The patient suffers from a lesion involving the

    (A) external iliac artery
    (B) popliteal artery or femoral artery distal to the origin of profunda femoris
    (C) profunda femoris
    (D) anterior tibial artery
    (E) posterior tibial artery

271. The cruciate anastomosis in the upper thigh is formed by the anastomosis of branches from all of the following arteries EXCEPT:

    (A) medial circumflex femoral artery
    (B) lateral circumflex femoral artery
    (C) superior gluteal artery
    (D) inferior gluteal artery
    (E) first perforating branch of the profunda femoris

272. The peroneal artery in the leg is a branch of the

    (A) femoral artery
    (B) profunda femoris
    (C) popliteal artery
    (D) anterior tibial artery
    (E) posterior tibial artery

The answers are: **269-D; 270-B; 271-C; 272-E.** Peripheral atherosclerotic occlusive disease is a condition in which medium or larger arteries of the extremities become occluded by plaques. The most common initial symptom of the disease in a lower limb is muscular pain or fatigue that occurs with exercise but abates with rest (intermittent claudication).

Occlusions in the external iliac (EI) artery diminish blood supply to almost all of the lower limb's muscles (Figs. 14 and 15), and thus produce exertion-dependent pain extending distally from the buttock. Occlusions in the femoral (F) artery immediately proximal to the origin of profunda femoris (PF) diminish blood supply to thigh and leg muscles, and thus produce exertion-dependent pain extending distally from the thigh. Occlusions in the popliteal (P) artery or femoral artery distal to the origin of profunda femoris produce exertion-dependent pain in the leg muscles. Progressive obstruction in the anterior or posterior tibial arteries does not diminish blood supply to the leg muscles, and thus does not produce intermittent claudication.

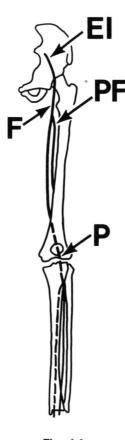

Fig. 14

In the buttock and back of the thigh, the superior and inferior gluteal (SG and IG) arteries, the medial and lateral circumflex femoral (MCF and LCF) arteries and the four perforating branches (PB) of the profunda femoris give rise to ascending and/or descending branches (Fig. 15). The anastomoses among these ascending and descending branches form a vertical chain of arteries in the posterior compartment of the thigh which extends from branches of the internal iliac artery in the buttock to genicular branches of the popliteal artery in the knee region. The cruciate anastomosis is one of the major anastomoses in this vertical chain. Under emergency conditions, the femoral artery can be ligated at any point along its course through the anterior compartment of the thigh without risking total loss of blood supply to the lower limb distal to the site of ligation because the vertical chain of anastomosed arteries in the posterior compartment of the thigh provides collateral circulation to the knee, leg and foot. The peroneal artery is a chief source of blood supply to the tissues of the lateral compartment of the leg. Peroneus longus and peroneus brevis are the only muscles in the lateral compartment of the leg.

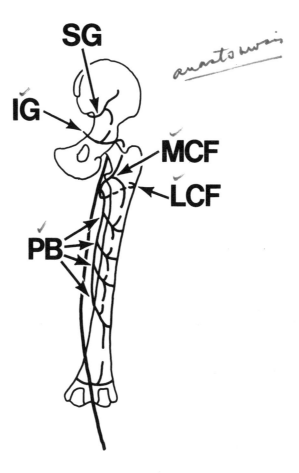

Fig. 15

Upon returning home after a 10-hour airline flight, a 55 year-old woman begins to experience pain in the calf of the left leg that intensifies upon standing or walking. Examination reveals tachycardia (pulse of 90), fever (100.5° F.) and cyanosis of the skin of the lower leg and foot.

273. The most likely diagnosis for the patient's calf pain is

    (A) an arterial aneurysm in the left lower limb
    (B) deep venous thrombosis of the left lower limb
    (C) incompetent valves in the deep veins of the left leg
    (D) Raynaud's disease
    (E) a tear of one of the heads of gastrocnemius in the left lower limb

274. What is a possible complication of the patient's disorder?

    (A) Carotid embolism
    (B) Subclavian embolism
    (C) Pulmonary embolism
    (D) Superior mesenteric embolism
    (E) Femoral embolism

275. Medium-sized and large veins of the lower limb may be palpated for tenderness in cases of suspected deep venous thrombosis. All of the following sites are appropriate for palpation of venous tenderness EXCEPT:

    (A) region deep to the medial aspect of the Achilles tendon
    (B) anterior region between the tibia and fibula in the lower leg
    (C) calf region overlying the soleus muscle
    (D) popliteal fossa
    (E) region immediately lateral to the pulsations of the femoral artery along the superior border of the femoral triangle

276. The great and small saphenous veins are the largest superficial veins of the lower limb. All of the following statements concerning the great and small saphenous veins are correct EXCEPT:

    (A) The great and small saphenous veins begin, respectively, as the medial and lateral extensions of the dorsal venous arch of the foot.
    (B) The small saphenous vein extends upward into the leg by passing behind the lateral malleolus.
    (C) The great saphenous vein extends upward into the leg by passing behind the medial malleolus.
    (D) The tributaries of the great saphenous vein drain all the superficial tissues of the lower limb except those of the lateral side of the foot and the posterolateral aspect of the leg.
    (E) The small saphenous vein commonly ends by uniting with the popliteal vein.

ANSWERS AND TUTORIAL FOR ITEMS 273-276

The answers are: **273-B; 274-C; 275-E; 276-C.** The patient suffers from deep venous thrombosis of the left lower limb. A venous thrombosis is an occlusion of a vein by a thrombus. In the extremities, the symptoms associated with acute venous thrombosis can vary from the absence of any symptoms to severe localized pain and evidence of systemic inflammation (such as fever, tachycardia and anxiety). Limited collateral venous drainage may produce cyanosis of the skin. The three major predisposing factors for venous thrombosis are venous stasis, alterations in the venous wall and abnormalities in the blood coagulation system.

An arterial embolism is the sudden obstruction of an artery by a clot or plug transported by blood flow from the heart or other blood vessel to the site of obstruction. A thrombus dislodged from a deep leg vein could be transported into the right atrium of the heart by consecutively passing through the popliteal vein, femoral vein, external iliac vein, common iliac vein and inferior vena cava. If the embolus passed from the right atrium into the right ventricle and were ejected into the pulmonary trunk, it would finally become lodged within one of the pulmonary arteries or its branches.

There are several sites in the lower limb at which deep veins can be palpated for tenderness from venous thrombosis. The region deep to the medial aspect of the Achilles tendon contains the posterior tibial vein. The anterior region between the tibia and fibula in the lower leg contains the anterior tibial vein. The intramuscular veins of the soleus are a common site of deep venous thrombosis in the lower limb and these veins can be compressed by palpation of the upper part of the calf of the leg. Deep palpation of the popliteal fossa puts pressure upon the popliteal vein. In the most superior part of the femoral triangle, the femoral vein can be compressed by palpation of the region immediately medial to the pulsations of the femoral artery. The great saphenous vein is frequently selected for intravenous administration of fluids and medications. It ascends from the foot into the leg by passing in front of (anterior to) the medial malleolus.

The parents of a 10 year-old boy report that their son has constant, intense left upper leg pain. The patient denies any injury to the left lower limb, but plays soccer daily. The patient's temperature is 101.0° F. The patient states that the upper leg pain started last evening and causes him to limp because bearing weight on the left leg intensifies the pain. Examination reveals warm, erythematous skin overlying the anterior aspect of the proximal end of the tibia. Deep palpation at this site elicits tenderness. Passive flexion and extension of the leg at the knee are to normal limits but uncomfortable.

277. The history and physical exam suggest that the patient has

   (A) Osgood-Schlatter disease
   (B) infrapatellar bursitis
   (C) osteochondritis dissecans of the knee
   (D) osteomyelitis of the tibia
   (E) septic arthritis of the knee joint

278. Which of the following groups of lymph nodes is commonly the first to receive lymph drained from the deep tissues of the leg?

   (A) Popliteal nodes
   (B) Vertical group of superficial inguinal nodes
   (C) Horizontal group of superficial inguinal nodes
   (D) Deep inguinal nodes
   (E) External iliac nodes

279. Which of the following groups of lymph nodes is commonly the first to receive lymph drained from the superficial tissues on the medial side of the heel?

   (A) Popliteal nodes
   (B) Vertical group of superficial inguinal nodes
   (C) Horizontal group of superficial inguinal nodes
   (D) Deep inguinal nodes
   (E) External iliac nodes

ANSWERS AND TUTORIAL FOR ITEMS 277-279

The answers are: **277-D; 278-A; 279-B.** The patient is suffering from osteomyelitis of the proximal tibia. An individual with osteomyelitis generally presents with constant and intense pain at the site of infection. If the septic bone is relatively superficial (such as the anterior aspect of the proximal end of the tibia), the overlying skin may be warm and deep palpation may elicit tenderness. Since any act which increases intraosseous pressure increases the pain, the patient in this case avoids weight-bearing on the afflicted lower limb. Osteomyelitis in children is frequently caused by the hematogenous seeding of bacteria in an injured metaphysis of one of the long bones of the extremities. Blood flow through metaphyseal blood vessels is normally sluggish, and thus trauma which partially occludes the metaphyseal vasculature predisposes the vasculature to the deposition and proliferation of blood-borne bacteria.

The lymphatics which drain the deep tissues of the leg and foot are afferent to either the popliteal nodes or deep inguinal nodes. The popliteal nodes should react first to the septic tibial metaphysis because of their proximity to the metaphysis. The superficial tissues of the lower limb whose venous drainage is collected by the great saphenous vein correspond to the superficial tissues whose lymphatics are afferent to the vertical group of superficial inguinal nodes. The tributaries of the great saphenous vein drain all the superficial tissues of the lower limb except those of the lateral side of the foot and the posterolateral aspect of the leg. The veins which drain the superficial tissues of the lateral side of the foot and the posterolateral aspect of the leg are tributaries of the small saphenous vein. The lymphatics emanating from these tissues are afferent to the popliteal nodes.

In the following items, match each bone with its image in the labeled lateral radiograph of the foot.

280. Talus

281. Navicular

282. Calcaneus

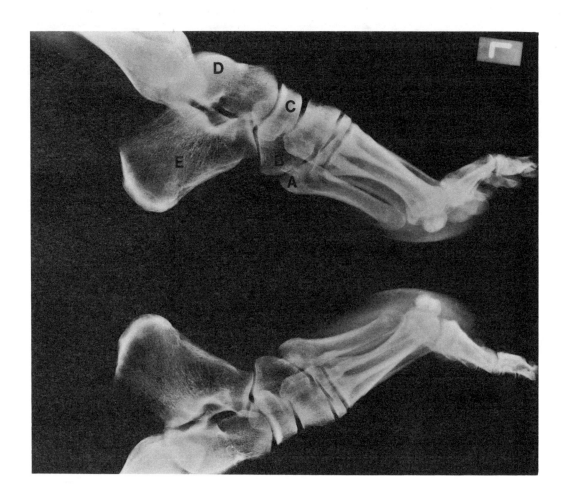

ANSWERS AND TUTORIAL FOR ITEMS 280-282

The answers are: **280-D; 281-C; 282-E.** The part of the bone labeled A is the base of the 5th metatarsal, and the tarsal bone labeled B is the cuboid. The talus articulates with the tibia and fibula in the ankle joint. The talus forms the keystone of the medial longitudinal arch of the foot. The calcaneus forms the heel of the foot.

Items 283-287

A 60 year-old man complaining of chest wall pain has a rash extending along the right 7th intercostal space. The rash consists of clear vesicles with an erythematous base. A diagnosis of Herpes zoster is made on the basis of the cutaneous distribution and appearance of the vesicular lesions.

283. The rash is the result of reactivation of a latent viral infection of the

    (A) intercostal arteries of the right 7th intercostal space
    (B) intercostal veins of the right 7th intercostal space
    (C) lymphatics of the right 7th intercostal space
    (D) intercostal muscles of the right 7th intercostal space
    (E) dorsal root ganglion of the right 7th intercostal nerve

284. Which rib's costal cartilage articulates with the sternum at the level of the sternal angle?

    (A) 1st
    (B) 2nd
    (C) 3rd
    (D) 4th
    (E) 5th

285. Which rib's costal cartilage is the lowest costal cartilage to contribute to the costal margin of the rib cage?

    (A) 8th
    (B) 9th
    (C) 10th
    (D) 11th
    (E) 12th

286. All of the following statements regarding the 7th intercostal nerve are correct EXCEPT:

    (A) Branches of the 7th intercostal nerve innervate the skin overlying the 7th intercostal space.
    (B) Branches of the 7th intercostal nerve innervate the parietal pleura underlying the 7th intercostal space.
    (C) Branches of the 7th intercostal nerve innervate the skin overlying the tip of the xiphoid process.
    (D) Branches of the 7th intercostal nerve innervate the external and internal intercostal muscles of the 7th intercostal space.
    (E) As the 7th intercostal nerve extends through its intercostal space, it lies close to the upper border of the 8th rib.

287. All of the following statements regarding the intercostal arteries and veins of the 7th intercostal space are correct **EXCEPT**:

    (A) The anterior intercostal arteries of the 7th intercostal space are branches of the musculophrenic artery.

    (B) The posterior intercostal artery of the 7th intercostal space is a branch of the descending thoracic aorta.

    (C) The intercostal veins of the 7th intercostal space drain anteriorly into the musculophrenic vein.

    (D) The intercostal veins of the 7th intercostal space drain posteriorly into the azygos system of veins.

    (E) As the intercostal arteries and veins extend through the 7th intercostal space, they lie superficial to the internal intercostal muscle.

ANSWERS AND TUTORIAL FOR ITEMS 283-287

The answers are: **283-E; 284-B; 285-C; 286-E; 287-E.** The vesicular lesions of Herpes zoster are characteristically distributed throughout the dermatome of the spinal nerve with the infected dorsal root ganglion. The sternal angle (SA) is the posterior angle between the manubrium and the body of the sternum at the manubriosternal joint (MJ) (Fig. 16). The manubriosternal joint can be easily palpated as a horizontal ridge in the midline of the anterior chest wall because the sternal angle is less than 180 degrees. The costal cartilage of the 2nd rib almost invariably articulates with the sternum at the level of the sternal angle, or manubriosternal joint (MJ) (Fig. 17). Palpation of the manubriosternal joint thus permits numerical identification of the ribs. The costal cartilages of the 7th to 10th ribs form, on each side, the costal margin (CM) of the rib cage. Each intercostal nerve innervates the intercostal muscles of its intercostal space and provides sensory innervation for the skin overlying the intercostal space and the parietal pleura underlying the intercostal space. Branches of the 7th intercostal nerve also provide sensory innervation for the skin overlying the most superior part of the anterolateral abdominal wall (including the skin overlying the xiphoid process) and the parietal peritoneum underlying this abdominal wall region.

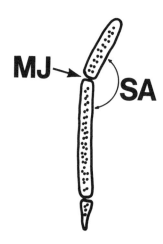

Fig. 16

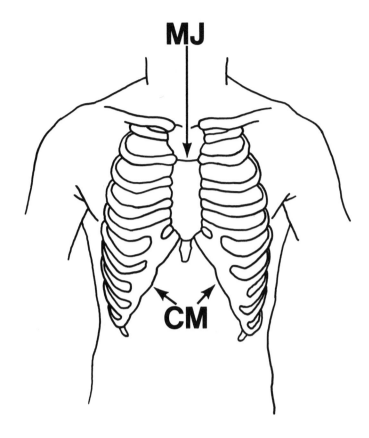

Fig. 17

Each intercostal nerve (N) extends through its intercostal space closely aligned with an intercostal artery (A) and intercostal vein (V) (Fig. 18). This neurovascular bundle lies partially under the cover of the costal groove (CG) of the upper rib and directly deep to the internal intercostal muscle (IIM).

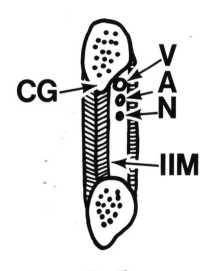

Fig. 18

The following items address information an examiner needs to know when percussing the lungs.

288. Which labeled area in the following drawing of the anterior chest wall marks the region where the apex of the right lung can be percussed?

B

apex above clavicle

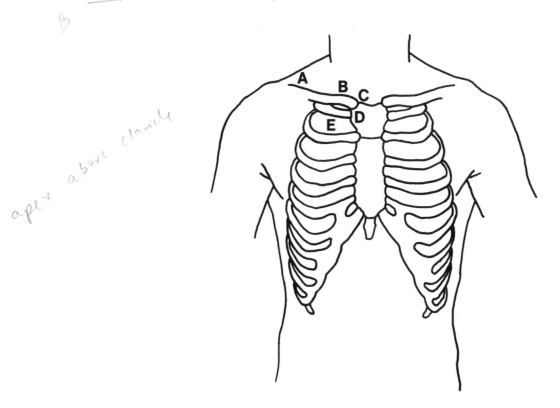

289. Which labeled area in the following drawing of the anterior chest wall is the lowest area that the upper lobe of the left lung can be percussed?

D

lowest area of upper lobe

oblique fissure ✓

of inferior lobe ✓

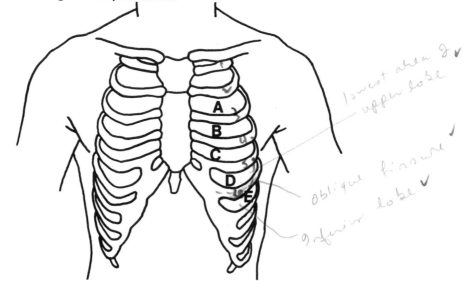

290. Which labeled area in the following drawing of the posterior chest wall is the lowest area that the upper lobe of the left lung can be percussed?

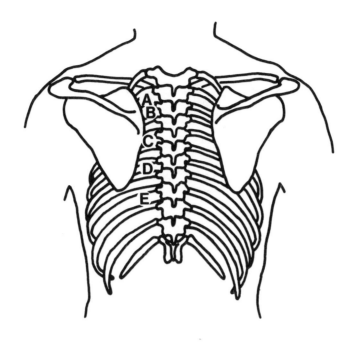

291. At mid-inspiration, which labeled level in the following drawing of the anterior chest wall marks the surface projection of the horizontal fissure of the right lung?

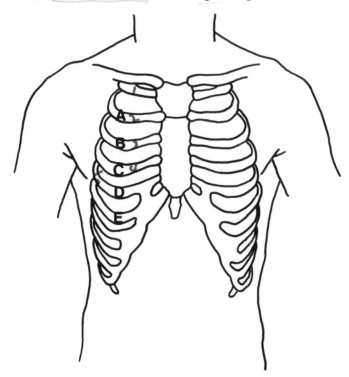

292. At full and deep inspiration, which labeled level in the following drawing of the posterior chest wall marks the level of the costodiaphragmatic margin of the right lung?

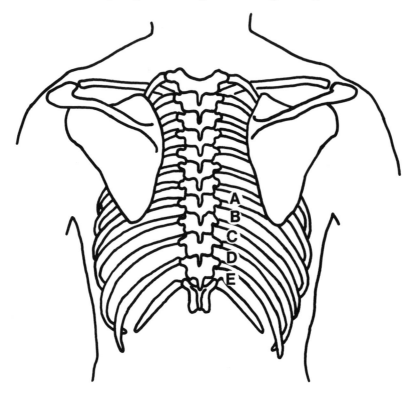

293. Which labeled level in the following drawing of the lateral chest wall marks the level of the costodiaphragmatic margin of the left pleural cavity along the midaxillary line?

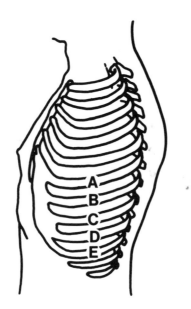

ANSWERS AND TUTORIAL FOR ITEMS 288-293

The answers are: **288-B; 289-D; 290-B; 291-C; 292-E; 293-E.** The bony and cartilaginous parts of the rib cage can serve to mark the surface projections of the pleural cavities and the lobes of the lungs. The most important surface relationships are as follows:

(1) The apex of each lung (A) projects above the medial third of the clavicle (C) (Fig. 19).

(2) For both lungs, the surface projection of the oblique fissure (OF) is an arc which begins posteriorly between the tips of the spinous processes of the 4th and 5th thoracic vertebrae, crosses the 5th and 6th ribs as it extends laterally to the midaxillary line, and then overlaps the lower border of the 6th rib to the costodiaphragmatic margin of the lung (Figs. 19 and 20).

(3) For the right lung, the surface projection of the horizontal fissure (HF) is an arc which overlaps the 4th rib from the midaxillary line to the lateral border of the sternum (Figs. 19 and 20).

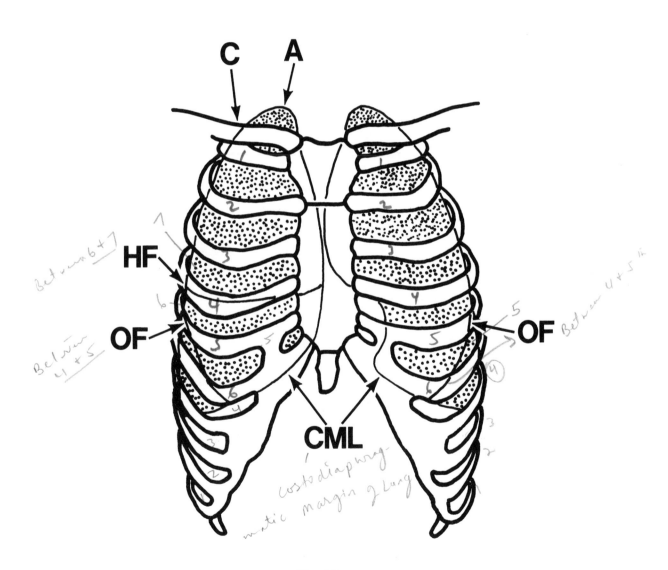

Fig. 19

127

The costodiaphragmatic margin of each pleural cavity lies at the level of the 8th rib at the midclavicular line, the 10th rib at the midaxillary line, and the 11th or 12th rib at the lateral border of the vertebral column. At mid-inspiration, the costodiaphragmatic margin of each lung (CML) lies 2 rib spaces above the costodiaphragmatic margin of its pleural cavity (Figs. 19 and 20). The costodiaphragmatic margin of each lung descends inferiorly to the costodiaphragmatic margin of its pleural cavity at full and deep inspiration.

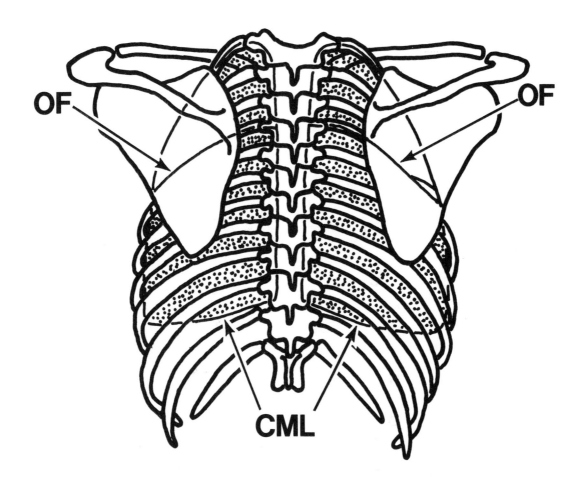

Fig. 20

Items 294-296

An asymptomatic, 22 year-old woman is found during a routine physical exam to have a mid-systolic click followed by a late systolic crescendo type murmur. Echocardiography shows a prolapsing posterior leaflet of the mitral valve.

294. Which of the labeled positions in the following drawing of the anterior chest wall is the best site for hearing the murmur produced by the prolapsing mitral valve?

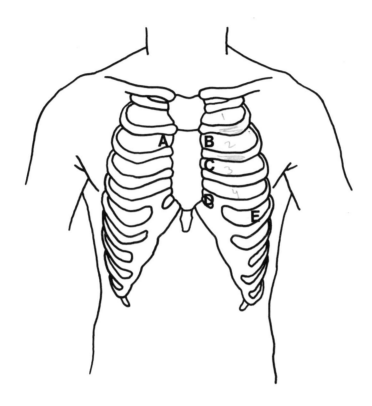

295. Certain procedures commonly move the late systolic murmur of a prolapsing mitral valve either toward or away from S1 (the first heart sound). All of the following statements concerning these procedures are correct **EXCEPT:**

    (A) Administration of a vasodilator should move the murmur away from S1.
    (B) Moving from a supine to a standing position should move the murmur toward S1.
    (C) Squatting should move the murmur away from S1.
    (D) Exertion of the Valsalva maneuver should move the murmur toward S1.
    (E) Making tight fists with the hands should move the murmur away from S1.

296. The first heart sound has two components because it is produced by the near-simultaneous closure of two valves. These two valves are the

   (A) tricuspid and pulmonary valves
   (B) mitral and aortic valves
   (C) aortic and pulmonary valves
   (D) tricuspid and mitral valves
   (E) pulmonary and mitral valves

ANSWERS AND TUTORIAL FOR ITEMS 294-296

The answers are: **294-E; 295-A; 296-D.** The best site for hearing the closure of the mitral valve is the site in the left 4th or 5th intercostal space where the apex of the heart beats against the anterior chest wall during systole. This site is generally also the best site for hearing the murmur produced by a prolapsing mitral valve. In patients with a prolapsing mitral valve, prolapse does not generally occur until mid-systole. It is believed that the mid-systolic click emanates from the prolapsed mitral valve leaflet or its chordae tendineae when they are acutely tensed at the moment of maximum prolapse. Regurgitation of blood from the left ventricle into the left atrium through the prolapsed mitral valve generates the late systolic murmur. Procedures which decrease systemic venous return to the heart (administration of a vasodilator, moving from a supine to a standing position, or exertion of the Valsalva maneuver) decrease left ventricular end-diastolic volume and thus move the click and murmur toward S1. By contrast, procedures which increase systemic venous return to the heart (squatting or isometric exercises) increase left ventricular end-diastolic volume and thus delay the onset of the click and murmur.

The beginning of systole during the cardiac cycle is marked by ventricular contraction concurring with atrial relaxation. These concurrent events on both sides of the heart promptly generate a blood pressure in each contracting ventricle which exceeds that in the adjoining, relaxing atrium. This pressure difference across each atrioventricular valve closes the valve shut. The mitral and tricuspid valves close shut almost simultaneously, producing audible vibrations in the blood confined to the ventricular chambers. Collectively, the vibrations form S1, the first heart sound (the 'lub' sound) of each heartbeat. The closure of the mitral valve slightly precedes the closure of the tricuspid valve.

Items 297-301

A 42 year-old man enters the emergency department complaining of dyspnea and severe chest pain. The physical findings suggest pericardial tamponade. An ECG showing total electrical alternans and an echocardiogram showing pericardial effusion confirm the diagnosis.

297. All of the following physical findings are commonly encountered in cases of pericardial tamponade **EXCEPT:**

    (A) shift of the right border of cardiac dullness to the right
    (B) sinus tachycardia
    (C) bilateral distension of the jugular veins
    (D) increased intensity of S1 and S2
    (E) decrease of the systolic blood pressure

298. Which jugular vein is commonly the best barometer of central venous pressure?

    (A) Right anterior jugular vein
    (B) Right external jugular vein
    (C) Right internal jugular vein
    (D) Left external jugular vein
    (E) Left internal jugular vein

299. All of the following statements concerning the jugular, subclavian and brachiocephalic veins are correct **EXCEPT:**

    (A) The anterior jugular vein ends via union with either the external jugular or subclavian vein.
    (B) The internal jugular vein ends at its union with the subclavian vein.
    (C) The union of the external jugular vein with the subclavian vein forms the brachiocephalic vein. *+ Internal jugular +*
    (D) The union of the left and right brachiocephalic veins forms the superior vena cava.
    (E) In the mediastinum, the right brachiocephalic vein is more vertical than the left brachiocephalic vein.

300. If the head, neck and trunk of a supine adult are elevated 30-45 degrees from the horizontal, what is the normal vertical distance above the level of the sternal angle that pulsatile activity will be observed in the jugular veins?

    (A) 0-1 cm
    (B) 2-3 cm
    (C) 4-5 cm
    (D) 6-7 cm
    (E) 8-9 cm

301. In a normal, healthy adult, the right border of the heart lies to the right of the right border of the sternum by

    (A) 0-1 cm
    (B) 2-3 cm
    (C) 4-5 cm
    (D) 6-7 cm
    (E) 8-9 cm

ANSWERS AND TUTORIAL FOR ITEMS 297-301

The answers are: **297-D; 298-C; 299-C; 300-B; 301-A.** Pericardial tamponade (accumulation of fluid within the pericardial cavity) broadens the region of cardiac dullness in the anterior chest wall, and thus shifts the right border of cardiac dullness to the right from its normal location at the right border of the sternum. The extracardiac pressure exerted by the pericardial fluid decreases intracardiac diastolic pressures, and thereby reduces venous return to the right atrium (as evidenced by bilateral distension of jugular veins) and systolic blood pressure. Sinus tachycardia occurs as a compensatory mechanism to maintain cardiac output. The pericardial fluid diminishes the intensity of S1 and S2.

The internal and external jugular (IJ and EJ) veins, subclavian (S) veins, brachiocephalic (B) veins and superior vena cava (SVC) are the major venous trunks which extend inferiorly through the neck and superior mediastinum to conduct blood into the right atrium of the heart (Fig. 21). When a normal adult is standing or seated upright, blood fills these venous trunks up to a level about 2-3 cm above the sternal angle. In other words, under these conditions, blood fills the superior vena cava, brachiocephalic veins and subclavian veins but only the lowest parts of the jugular veins. The height to which the jugular veins are filled with blood is proportional to right atrial pressure. Accordingly, the jugular veins can serve as manometers of right atrial pressure. Right atrial pressure is frequently called central venous pressure, since the blood pressure of the right atrium approximates that of the large systemic veins converging upon the right atrium. The right internal jugular vein is the most appropriate jugular vein to select for the measurement of central venous pressure. This is because it is the only jugular vein which forms a straight-lined venous trunk with the right brachiocephalic vein and superior vena cava (Fig. 21). The angular union of the left internal jugular vein with the left subclavian vein and the angular unions of the external jugular veins with the subclavian veins make the blood heights in these jugular veins less reliable monitors of central venous pressure.

During each heartbeat, there are three transient increases in right atrial pressure. The first pulsatile increase (the a wave) occurs when contraction of the atrial musculature increases atrial pressure. The second increase (the c wave) occurs during early systole, when the increasing pressure in the right ventricle bulges the tricuspid valve into the right atrium. The third pulsatile increase (the v wave) occurs as blood flowing into the right atrium during late systole begins to bulge the heart chamber. These pulsatile increases are transmitted in a retrograde fashion through the blood in the internal and external jugular veins. When these jugular venous pulses reach the meniscus (curved, upper surface) of the blood in each jugular vein, they produce fluctuations in the level of the meniscus, and these meniscal fluctuations, in turn, produce up-and-down movements of the overlying skin. These skin movements over each jugular vein are generally the best indicator of the height to which blood fills the vein.

Inspection of jugular pulses should begin with the patient in the supine position. In a normal individual, the jugular veins will be distended because the jugular veins and the right atrium now are all at about the same level. The patient's head, neck and trunk are next elevated above the horizontal plane sufficiently to lower the height of the jugular pulses to a level which is below the angle of the mandible but above the clavicle (a 30-45 degree elevation is generally sufficient). In a normal adult, pulsatile activity will be visible near the lower ends of the jugular veins, commonly up to a vertical distance of 2-3 cm above the level of the sternal angle. This is because the average central venous pressure in a normal individual is about 7-8 cm water, and the center of the right atrium is about 5 cm below the level of the sternal angle in an average adult.

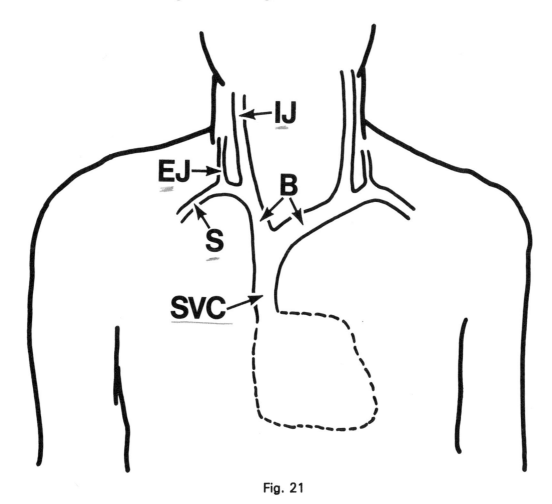

Fig. 21

Items 302-305

A 63 year-old woman enters the emergency department complaining of exertional chest pain and exertional syncope. An ejection murmur and other findings suggest severe aortic stenosis. An ECG showing left ventricular hypertrophy and an echocardiogram showing a thickened, calcified aortic valve confirm the diagnosis.

302. All of the following physical findings are commonly encountered in cases of severe aortic stenosis **EXCEPT:**

    (A) diminished intensity of S2
    (B) slow-rising carotid arterial upstrokes
    (C) radiation of the ejection murmur to the carotid arteries
    (D) shift of the apical thrust upward and laterally
    (E) marked precordial apical thrust

303. The second heart sound has two components because it is produced by the near-simultaneous closure of two valves. These two valves are the

    (A) aortic and pulmonary valves
    (B) tricuspid and mitral valves
    (C) aortic and mitral valves
    (D) pulmonary and mitral valves
    (E) aortic and tricuspid valves

304. The apex beat [apical thrust, or PMI (point of maximum impulse)] is normally found

    (A) 3-5 cm to the left of the midsternal line in the 5th intercostal space
    (B) 5-7 cm to the left of the midsternal line in the 5th intercostal space
    (C) 7-9 cm to the left of the midsternal line in the 5th intercostal space
    (D) 5-7 cm to the left of the midsternal line in the 6th intercostal space
    (E) 7-9 cm to the left of the midsternal line in the 6th intercostal space

305. Which of the labeled positions in the following drawing of the anterior chest wall is the best site for hearing the closure of the aortic valve?

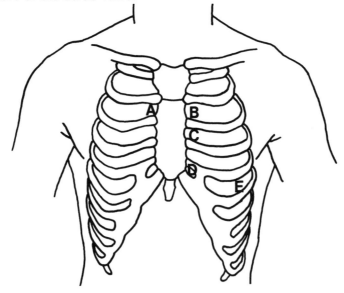

ANSWERS AND TUTORIAL FOR ITEMS 302-305

The answers are: **302-D; 303-A; 304-B; 305-A.** In this case, aortic stenosis is the result of progressive calcification of the aortic valve leaflets. Left ventricular hypertrophy has served to maintain left ventricular output. A stenosed aortic valve restricts and disturbs left ventricular outflow during systole. The restriction prolongs ejection and thus delays the attainment of peak carotid pressure. The disturbance of left ventricular outflow produces an ejection murmur that is transmitted in an antegrade fashion to the carotid arteries. The intensity of the sound produced by the closure of the aortic valve at the end of systole is diminished because the valve leaflets are thick and stiff. The contraction of the hypertrophic left ventricle produces the marked precordial apical thrust. Left ventricular hypertrophy shifts the apex of the heart, and thus the PMI, downward and laterally. The exertional angina is believed to be due to insufficiency of blood supply to the hypertrophic left ventricular muscle mass. Such insufficiency can occur with even unobstructed coronary arteries. The exertional syncope is due to inadequate cerebral perfusion.

Toward the end of systole, after the ventricles have ejected most of their blood, there begins a retrograde flow of blood from the pulmonary trunk and aorta into the ventricles (as a consequence of the blood pressure in each arterial trunk being greater than that in the ventricle). This retrograde blood flow snaps the pulmonary and aortic valves shut. Their near-simultaneous closure generates audible vibrations in the blood borne by both arterial trunks. Collectively, these vibrations produce S2, the second heart sound (the 'dup' sound) of the heartbeat. The closure of the aortic valve precedes the closure of the pulmonary valve during inspiration.

Items 306-308

A 34 year-old woman reports recent exertional chest pain. The physical and radiographic findings suggest advanced primary pulmonary hypertension.

306. All of the following findings are consistent with this diagnosis **EXCEPT**:

    (A) a palpable parasternal heave
    (B) diminished intensity of the pulmonic component of S2
    (C) a PA chest film showing enlarged pulmonary arteries
    (D) shift of the apical thrust upward and laterally
    (E) increased jugular venous pressure

307. Which of the labeled positions in the following drawing of the anterior chest wall is the best site for hearing the closure of the pulmonary valve?

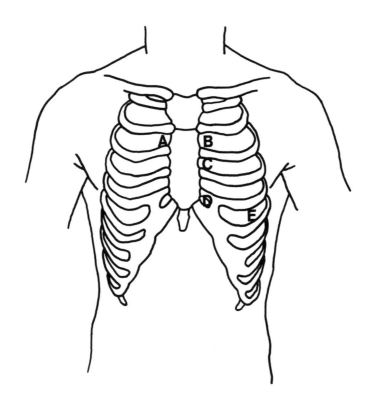

308. All of the following statements concerning the bronchial arteries and veins of the lungs are correct EXCEPT:

    (A) The root of the lung transmits the bronchial arteries into the lung and the bronchial vein out from the lung.
    (B) The bronchial vein drains the capillary beds of the conducting airways of the lung.
    (C) The azygos system of veins commonly drains the bronchial veins.
    (D) The bronchial arteries supply the conducting airways of the lung.
    (E) The bronchial arteries are branches of the aortic arch.

ANSWERS AND TUTORIAL FOR ITEMS 306-308

The answers are: **306-B; 307-B; 308-E.** In primary pulmonary hypertension, the resistance to pulmonary blood flow increases due to smooth muscle hypertrophy and intimal proliferation of the pulmonary arteries and arterioles. Right ventricular output is maintained initially by right ventricular hypertrophy. Contraction of the hypertrophic right ventricle produces a palpable, parasternal heave. Right ventricular hypertrophy also shifts the apex of the heart, and thus the PMI, upward and laterally. The significant elevation of pulmonary arterial pressure enlarges the pulmonary arteries and causes a more forceful closure of the pulmonic valve at the end of systole thus producing a louder pulmonic component in S2. Inadequacy of right ventricular function eventually leads to venous congestion and a rise in central venous pressure, evidence of which is provided by elevated jugular venous pressure. Distension of the pulmonary arteries or their major branches causes the patient's exertional chest pain.

    Each lung is supplied by one or two bronchial arteries. These arteries arise from either the descending thoracic aorta or the upper posterior intercostal arteries.

A 37 year-old woman visits a cardiologist because her medication does not manage her unstable angina. The cardiologist recommends coronary arteriography to evaluate the need for revascularization. In the following items, match each coronary artery or arterial branch with its image in the labeled drawing below. The drawing shows a view of the sternocostal surface of the heart.

309. Left anterior descending artery (anterior interventricular branch of the left coronary artery)

310. Right coronary artery

311. Posterior descending artery (posterior interventricular branch of the right coronary artery)

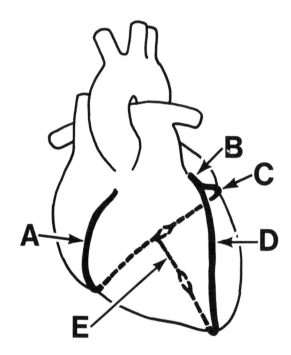

ANSWERS AND TUTORIAL FOR ITEMS 309-311

The answers are: **309-D; 310-A; 311-E.** The artery labeled B is the left coronary artery, and the artery labeled C is the circumflex artery. The left anterior descending artery descends in the anterior interventricular groove on the sternocostal surface of the heart. Its branches supply both ventricles and the interventricular septum. The right coronary artery extends along the atrioventricular groove on the sternocostal and diaphragmatic surfaces of the heart. In 90% of individuals, the right coronary artery extends along the posterior interventricular groove to become the posterior descending artery. The right coronary artery supplies all of the right atrium, much of the right ventricle and parts of the left atrium and left ventricle.

In the following items, match each cardiac vein with its image in the labeled drawing below. The drawing shows a view of the sternocostal surface of the heart.

312. Middle cardiac vein

313. Coronary sinus

314. Great cardiac vein

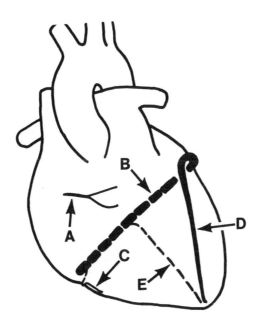

ANSWERS AND TUTORIAL FOR ITEMS 312-314

The answers are: **312-E; 313-B; 314-D**. A is the anterior cardiac vein, and C is the small cardiac vein. The great cardiac vein ascends in the anterior interventricular groove and then turns to the left to extend along the atrioventricular groove on the sternocostal surface of the heart. At the left border of the heart, the great cardiac vein gives rises to the coronary sinus. The coronary sinus extends to the right along the atrioventricular groove on the diaphragmatic surface of the heart. The coronary sinus ends at its opening into the posterior floor region of the right atrium. The middle cardiac vein ascends in the posterior interventricular groove on the diaphragmatic surface of the heart, and joins the coronary sinus at the crux. The crux is the point where the atrioventricular groove meets the posterior interventricular groove.

In the following items, match each structure with an image of one of its borders in the labeled PA radiograph of the chest.

315. Aortic arch (aortic knob)

316. Pulmonary trunk

317. Left ventricle

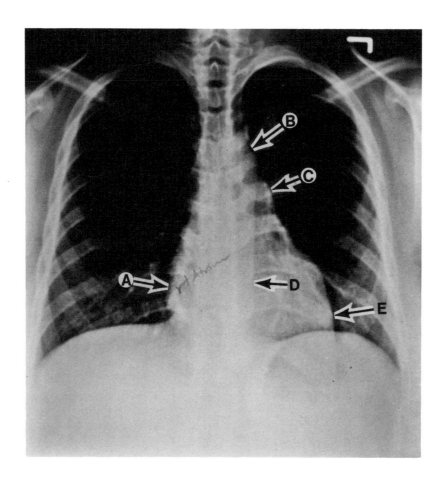

ANSWERS FOR ITEMS ITEMS 315-317

The answers are: **315-B; 316-C; 317-E.** The structure labeled A is the right border of the right atrium, and the structure labeled D is the left border of the descending thoracic aorta.

The three CT scans on the following page are adjacent 10 mm-thick thoracic scans. In the following items, match each structure with its image in the labeled CT scan.   *Pg 142*

318. Right brachiocephalic vein

319. Brachiocephalic trunk

320. Left common carotid artery

ANSWERS FOR ITEMS 318-320

The answers are: **318-A; 319-C; 320-D.** The structure labeled B is the left brachiocephalic vein, and the structure labeled E is the left subclavian artery.

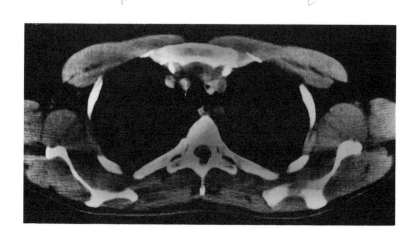

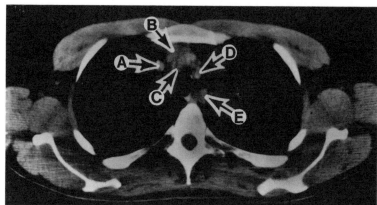

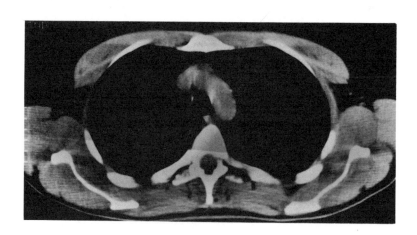

The three CT scans on the following page are adjacent 10 mm-thick thoracic scans. In the following items, match each structure with its image in the labeled CT scan.

321. Pulmonary trunk

322. Esophagus

323. Superior vena cava

ANSWERS FOR ITEMS 321-323

The answers are: **321-C; 322-D; 323-A.** The structure labeled B is the ascending aorta, and the structure labeled E is the descending thoracic aorta.

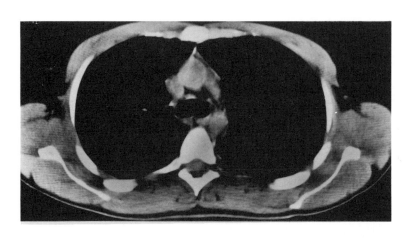

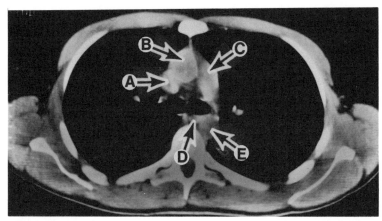

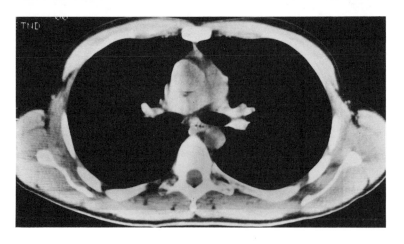

144

Items 324-330

A helpful way to understand the anatomy of the abdomen and pelvis is to visualize in the order shown the placement of the following five groups of organs:

[1] the retroperitoneal viscera of the abdomen,
[2] the secondarily retroperitoneal viscera of the abdomen,
[3] the intraperitoneal segments of the small and large intestines in the mid and lower abdomen,
[4] the liver and gallbladder and the intraperitoneal viscera in the upper abdomen, and
[5] the pelvic viscera.

The following items review the peritoneal relationships among abdominal and pelvic viscera.

324. All of the following viscera are retroperitoneal **EXCEPT:**

  (A) abdominal aorta
  (B) spleen
  (C) right kidney
  (D) left adrenal gland
  (E) inferior vena cava

325. All of the following viscera are secondarily retroperitoneal **EXCEPT:**

  (A) ascending colon
  (B) head of the pancreas
  (C) body of the pancreas
  (D) tail of the pancreas
  (E) descending colon

326. All of the following segments of the small and large intestines are intraperitoneal **EXCEPT:**

  (A) 2nd, 3rd and 4th parts of the duodenum
  (B) jejunum
  (C) ileum
  (D) transverse colon
  (E) sigmoid colon

327. All of the following regions of the abdominal and pelvic cavities are located in the greater sac **EXCEPT:**

  (A) region anterior to the mesentery of the small intestine
  (B) left lateral paracolic gutter
  (C) hepatorenal recess (Morrison's pouch)
  (D) region posterior to the stomach
  (E) rectouterine pouch (pouch of Douglas)

328. All of the following statements regarding the epiploic foramen are correct EXCEPT:

(A) The epiploic foramen is a passageway between the greater and lesser sacs of the peritoneal cavity.
(B) A part of the caudate lobe of the liver borders the foramen superiorly.
(C) The proximal half of the 1st part of the duodenum borders the foramen inferiorly.
(D) The portal vein borders the foramen posteriorly.
(E) The free right margin of the lesser omentum borders the foramen anteriorly.

329. All of the following peritoneal ligaments border a part of the lesser sac EXCEPT:

(A) lesser omentum
(B) lienorenal ligament
(C) gastrophrenic ligament
(D) gastrocolic ligament
(E) falciform ligament

330. All of the following statements concerning peritoneal relationships in the female pelvis are correct EXCEPT:

(A) The fundus of the uterus is covered with peritoneum.
(B) In the midline region, the peritoneum which covers the anterior surface of the middle third of the rectum is continuous anteriorly with the peritoneum that covers the uppermost posterior surface of the vagina.
(C) In the midline region, the peritoneum which covers the anterior surface of the uterus is continuous anteriorly with the peritoneum that covers the uppermost anterior surface of the vagina.
(D) The superior surface of the urinary bladder is covered with peritoneum.
(E) The ovary is attached by the mesovarium to the posterior peritoneal layer of the broad ligament of the uterus.

ANSWERS AND TUTORIAL FOR ITEMS 324-330

The answers are: **324-B; 325-D; 326-A; 327-D; 328-D; 329-E; 330-C.** The retroperitoneal viscera of the abdomen include the abdominal aorta, inferior vena cava, kidneys, adrenal glands and ureters. The spleen is intraperitoneal. The secondarily retroperitoneal viscera of the abdomen include the head, neck and body of the pancreas; the distal half of the 1st part of the duodenum and the 2nd, 3rd and 4th parts of the duodenum; the ascending colon with the hepatic flexure at the upper end; and the descending colon with the splenic flexure at the upper end. The tail of the pancreas is intraperitoneal, lying sandwiched between the two peritoneal layers of the lienorenal ligament. The lesser sac of the peritoneal cavity is sealed off superiorly by the diaphragm, anteriorly by the lesser omentum, stomach and gastrocolic ligament, inferiorly by the transverse colon and transverse mesocolon, posteriorly by the upper posterior abdominal wall and on the left by the gastrophrenic, gastrosplenic and lienorenal ligaments. On the right, the lesser sac communicates with the greater sac via the epiploic foramen, which has four borders: superiorly the caudate process of the caudate lobe of the liver, anteriorly the free right margin of the lesser omentum (this margin transmits the portal vein, hepatic artery proper and bile duct), inferiorly the proximal half of the 1st part of the duodenum, and posteriorly the inferior vena cava. In the midline region, the peritoneum which covers the anterior surface of the uterus (U) is continuous anteriorly with the peritoneum that covers the superior surface of the urinary bladder (UB) (Fig. 22).

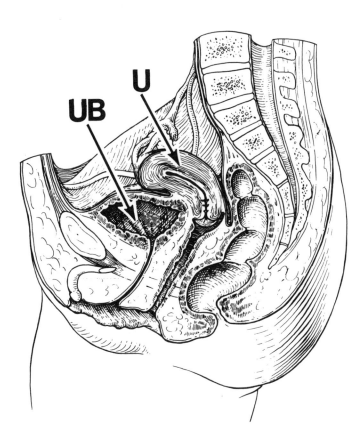

Fig. 22

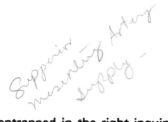

Items 331-334

A loop of ileum becomes entrapped in the right inguinal canal of a 10 year-old boy who has had a patent processus vaginalis here.

331. If pain fibers in the entrapped ileal loop produce visceral pain, in which abdominal region will the patient most likely feel the visceral pain?

    (A) Epigastric region
    (B) Umbilical region
    (C) Hypogastric region
    (D) Right inguinal region
    (E) Right lumbar region

332. Sensory fibers from the jejunum, ileum and cecum enter the spinal cord at spinal cord segment levels

    (A) C3, C4 and C5
    (B) T5, T6, T7, T8 and T9
    (C) T8, T9, T10, T11 and T12
    (D) T10, T11, T12, L1 and L2
    (E) L1, L2, S2, S3 and S4

333. All of the following statements concerning abdominal hernias are correct **EXCEPT:**

    (A) The neck of the sac of a femoral hernia always lies immediately medial and superior to the pubic tubercle.
    (B) The neck of the sac of an indirect inguinal hernia always lies lateral to the inferior epigastric artery.
    (C) The neck of the sac of a direct inguinal hernia always lies medial to the origin of the inferior epigastric artery.
    (D) The sac of a hernia consists of an outpouching of parietal peritoneum.
    (E) The neck of the sac of a hernia is the proximal end of the sac.

334. All of the following statements concerning the inguinal canal are correct **EXCEPT:**

    (A) The external spermatic fascia is continuous with the aponeurotic tendon of the external oblique muscle at the superficial inguinal ring.
    (B) The internal spermatic fascia is continuous with the transversalis fascia at the deep inguinal ring.
    (C) The aponeurotic tendon of the external oblique muscle forms most of the anterior wall of the inguinal canal.
    (D) The inguinal ligament forms the floor of the inguinal canal.
    (E) Transversalis fascia forms most of the roof of the inguinal canal.

The answers are: **331-B; 332-C; 333-A; 334-E.** Abdominal visceral pain is dull and sickening pain. It is poorly localized to one or more of the midline abdominal regions (epigastric, umbilical or hypogastric). Visceral pain is produced by the stimulation of visceral pain fibers. Visceral pain fibers are sensitive to acute stretching and anoxia. Disease or injury of the abdominal viscera supplied by the superior mesenteric artery produces visceral pain that is most commonly localized to the umbilical region. The greater, lesser and least splanchnic nerves provide almost all the preganglionic sympathetic innervation for the viscera supplied by the superior mesenteric artery. The sensory fibers that innervate the viscera supplied by the superior mesenteric artery enter the spinal cord at those spinal cord segments that provide preganglionic sympathetic innervation for the viscera (specifically, the sensory fibers enter at the T8-T12 levels).

An indirect inguinal hernia is a hernia whose contents protrude into the inguinal canal through the deep inguinal ring (DIR) (Fig. 23). The neck of the sac of an indirect inguinal hernia thus always lies lateral to the inferior epigastric artery (IEA). An indirect inguinal hernia almost always occurs in an individual with a patent processus vaginalis. A direct inguinal hernia is a hernia whose contents protrude into the inguinal canal through a distention of its posterior wall. The neck of the sac of a direct inguinal hernia always lies medial to the origin of the inferior epigastric artery and within the inguinal, or Hesselbach's, triangle. The inguinal triangle is the area within the inguinal region bounded by the inferior epigastric artery (IEA) laterally, the lateral edge of rectus abdominis (RA) medially, and the inguinal ligament (IL) inferiorly (Fig. 23). A femoral hernia is a hernia whose contents protrude through the femoral ring (FR) into the femoral canal (Fig. 23). The neck of the sac of a femoral hernia always lies immediately lateral and inferior to the pubic tubercle (PT). The lower free borders of internal oblique and transversus abdominis form the roof of the inguinal canal.

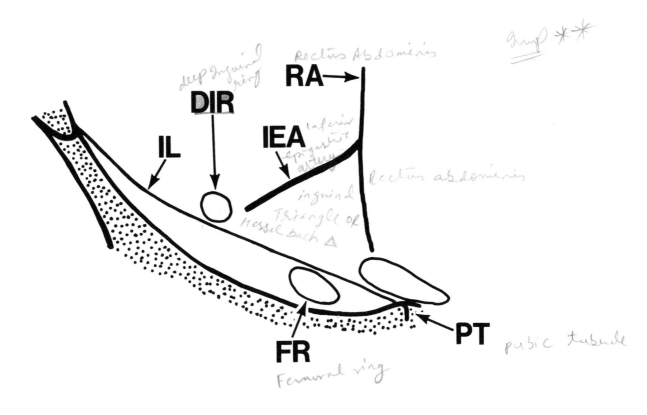

Fig. 23

Items 335-341

A 49 year-old woman reports upper abdominal pain of 5 days duration. The history and physical exam reveal pain typical of biliary colic, intermittent fevers with chills, yellowish sclera and oral mucosa, a tender gallbladder, and an enlarged and tender liver. Ultrasound examination of the patient's biliary tract shows gallstones in the lumen of the gallbladder, dilated intrahepatic biliary ducts and a gallstone lodged in the lower part of the common bile duct. The history, physical exam and radiological findings suggest ascending cholangitis (ascending sepsis of the extrahepatic and intrahepatic biliary ducts and the gallbladder).

335. Visceral pain from the biliary ducts is most often felt in the

(A) epigastric region
(B) umbilical region
(C) hypogastric region
(D) right hypochondriac region
(E) right lumbar region

336. Disease of the biliary ducts may produce referred pain in all of the following cutaneous regions **EXCEPT:**

(A) point of the shoulder
(B) right upper quadrant of the anterolateral abdominal wall
(C) anterior chest wall overlying the 6th, 7th and 8th intercostal spaces
(D) posterior chest wall overlying the inferior angle of the scapula
(E) posterior chest wall overlying the medial end of the spine of the scapula

337. Which set of spinal nerves innervates the skin in the center of the right upper quadrant of the anterolateral abdominal wall?

(A) T4, T5 and T6 on the right side
(B) T7, T8 and T9 on the right side
(C) T10, T11 and T12 on the right side
(D) L1 and L2 on the right side
(E) L3 and L4 on the right side

338. Percussion of the anterior chest and abdominal walls can be used to assess the height of the liver along the right midclavicular line. The normal range for the height of the liver in an adult is

(A) 3-6 cm
(B) 6-9 cm
(C) 9-12 cm
(D) 3-9 cm
(E) 6-12 cm

339. All of the following statements concerning the blood supply and venous drainage of the liver are correct **EXCEPT:**

(A) The right hepatic artery supplies most of the caudate lobe of the liver. *but not caudate Area*
(B) The left hepatic artery supplies the quadrate lobe of the liver, *caudate, lt hepatic lobe*
(C) The hepatic veins drain the blood conducted to the liver by the portal vein.
(D) The hepatic veins drain the blood conducted to the liver by the left and right hepatic arteries.
(E) The hepatic veins are tributaries of the inferior vena cava.

340. All of the following statements concerning arteries that arise directly or indirectly from the celiac artery are correct **EXCEPT:**

(A) Branches of the left gastric artery supply the stomach along the upper part of its lesser curvature.
(B) Branches of the splenic artery supply the body and tail of the pancreas.
(C) The left gastroepiploic artery arises from the splenic artery.
(D) The gastroduodenal artery lies directly anterior to the first part of the duodenum.
(E) Branches of the right gastroepiploic artery supply the stomach along the lower part of its greater curvature.

341. Portal-systemic anastomoses occur in all of the following regions **EXCEPT:**

(A) border region between the middle and lower thirds of the esophagus
(B) border region between the upper and lower halves of the anal canal
(C) border region between the upper and lower halves of the ureter
(D) paraumbilical region of the anterolateral abdominal wall
(E) retroperitoneal region deep to the posterior abdominal wall

ANSWERS AND TUTORIAL FOR ITEMS 335-341

The answers are: **335-A; 336-E; 337-B; 338-E; 339-A; 340-D; 341-C.** Disease or injury of the liver, gallbladder and other viscera supplied by the celiac artery produces visceral pain that is most commonly localized to the epigastric region. The left and right greater splanchnic nerves provide almost all the preganglionic sympathetic innervation for the viscera supplied by the celiac artery. Most of the sensory fibers that innervate the viscera supplied by the celiac artery enter the spinal cord at those spinal cord segments that give rise to the preganglionic sympathetic fibers for the greater splanchnic nerves (specifically, the sensory fibers enter at the T5-T9 levels). Some sensory fibers from the liver, gallbladder and the extrahepatic biliary ducts enter the spinal cord at the C3-C5 segments. The right phrenic nerve transmits these fibers to their cell bodies in the dorsal root ganglia of spinal nerves C3, C4 and C5. Disease or injury of any of the viscera supplied by the celiac artery may refer pain to parts of the T5-T9 dermatomes. Of the last four cutaneous regions listed in item 336, the right upper quadrant of the anterolateral abdominal wall represents part of the T7-T10 dermatomes, the anterior chest wall overlying the 6th, 7th and 8th intercostal spaces represents part of the T6-T8 dermatomes, the posterior chest wall overlying the inferior angle of the scapula represents part of the T7 dermatome, and the posterior chest wall overlying the medial end of the spine of the scapula represents part of the T3 dermatome. Disease or injury of the liver, gallbladder and extrahepatic ducts may also refer pain to the shoulder (as it represents part of the C3-C5 dermatomes).

Percussion can be used to assess the size of the liver. As the patient lies supine and holds the breath at full expiration, the examiner percusses the anterior chest wall and anterolateral abdominal wall from the right 2nd intercostal space downward along the right midclavicular line. Normally, percussion resonance (due to percussion of the right lung) is encountered down to the highest level (which is typically that of the 4th intercostal space) at which the liver crosses the right midclavicular line. A thin wedge of the right lung's lower lobe anteriorly covers the diaphragm and the underlying liver down to the level of the 6th rib. Percussion of both the right lung and the liver between the levels of the 4th intercostal space and the 6th rib produces a zone of percussion dullness called hepatic dullness. A zone of percussion flatness called hepatic flatness is encountered from the level of the 6th rib down to that of the liver's anteroinferior margin. Percussion of bowel segments inferior to the liver produce percussion dullness, resonance or tympany below the zone of hepatic flatness. The combined heights of the zones of hepatic dullness and hepatic flatness are a measure of the size of the liver. The normal range for the combined adult heights is 6-12 cm. The presence of gas-filled bowel segments immediately posterior to the lower part of the liver's visceral surface can obscure determination of the lower limit of the zone of hepatic flatness and lead to a faulty underestimate of liver size.

The left hepatic artery supplies the left hepatic lobe, quadrate lobe and caudate lobe except for the caudate process. The right hepatic artery supplies the rest of the hepatic parenchyma. Fig. 24 shows the relationships of the left gastric artery (LGA), splenic artery (SA), left gastroepiploic artery (LGEA), gastroduodenal artery (GA) and right gastroepiploic artery (RGEA) to the stomach (S), pancreas (P) and 1st part of the duodenum (D). The gastroduodenal artery passes behind the first part of the duodenum before giving rise to the right gastroepiploic and superior pancreaticoduodenal arteries. Portal-systemic anastomoses are venous anastomoses between veins that are tributaries of the portal vein and veins that are tributaries of the superior vena cava or inferior vena cava. In the border region between the middle and lower thirds of the esophagus, tributaries of the left gastric vein (which drains toward the portal vein) anastomose with tributaries of the azygos system of veins (which drains toward the superior vena cava). In the border region between the upper and lower halves of the anal canal, tributaries of the superior rectal vein (which drains toward the portal vein) anastomose with tributaries of the inferior rectal vein (which drains toward the inferior vena cava). In the paraumbilical region, tributaries of the paraumbilical veins (which drain toward the portal vein) anastomose with tributaries of the lumbar veins (which drain toward the inferior vena cava). In the retroperitoneal region deep to the posterior abdominal wall, tributaries of the splenic vein (which drains toward the portal vein) anastomose with tributaries of the left renal vein (which drains toward the inferior vena cava).

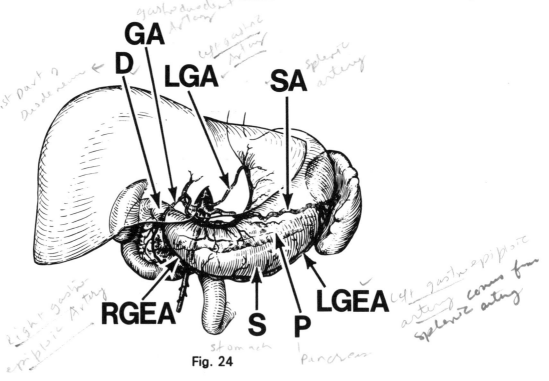

Fig. 24

152

Items 342-344

A 16-year old girl complains of abdominal pain, low grade fever (101.0° F.), nausea and anorexia during the last 12 hours. A tentative diagnosis of appendicitis is made on the basis of the history and physical exam.

342. Visceral pain from an inflamed appendix is most often felt in the

      (A) epigastric and/or umbilical regions
      (B) hypogastric region
      (C) right hypochondriac region
      (D) right lumbar region
      (E) right inguinal region

343. Assume that the appendix in the patient lies anterior to the terminal ileum and that inflammation has extended to the region of the anterolateral abdominal wall in contact with the appendix. Physical examination is likely to show all of the following findings **EXCEPT**:

      (A) deep tenderness in the right lower quadrant (RLQ) upon deep palpation of the RLQ
      (B) deep tenderness in the RLQ upon deep palpation of the left lower quadrant (LLQ)
      (C) pain in the abdominal wall of the RLQ when the approximated fingers of the examiner's hand are pressed into the abdominal wall of the RLQ and then suddenly withdrawn
      (D) pain in the abdominal wall of the RLQ when the approximated fingers of the examiner's hand are pressed into the abdominal wall of the LLQ and then suddenly withdrawn
      (E) right-sided rectal tenderness when right-sided pressure is applied to the rectum during a rectal exam

344. Assume that the appendix in the patient lies directly posterior to the cecum and that inflammation has extended to the region of the posterior abdominal wall in contact with the appendix. Physical examination is likely to show

      (A) pain upon abduction of the right thigh against resistance
      (B) pain upon flexion of the right thigh against resistance
      (C) pain upon adduction of the right thigh against resistance
      (D) pain upon passive internal rotation of the right thigh
      (E) pain upon passive external rotation of the left thigh

ANSWERS AND TUTORIAL FOR ITEMS 342-344

The answers are: **342-A; 343-E; 344-B.** Disease or injury of the appendix and other abdominal viscera supplied by the superior mesenteric artery produces visceral pain that is most commonly localized to the umbilical region. However, an inflamed appendix may elicit painful, contractive spasm of the pyloric sphincter. Stimulation of visceral pain fibers in the pyloric region of the stomach produces visceral pain that is most commonly localized to the epigastric region. Accordingly, appendicitis may produce visceral pain localized to the epigastric and/or umbilical regions. Localized deep tenderness in the RLQ is a common symptom of appendicitis prior to perforation.

The deep tenderness is either wholly or in part visceral pain that emanates from the appendix upon the application of external pressure and thus the site of deep tenderness always corresponds to the location of the appendix. Deep palpation of either the RLQ or LLQ may produce sufficient external pressure upon the inflamed appendix to elicit deep tenderness in the RLQ, particularly if the appendix lies within the right lateral paracolic gutter (position A in Fig. 25), on the parietal peritoneum overlying the iliacus muscle in the iliac fossa (position B) or anterior to the terminal ileum (position C). If the appendix lies posterior to the cecum (position D) or draped over the right lateral wall of the pelvis (position E), deep palpation of the lower quadrants may not produce sufficient external pressure upon the inflamed appendix to elicit deep tenderness in the RLQ.

Inflamed parietal peritoneum is exquisitely sensitive to sudden changes in tension. Such sensitivity in the parietal peritoneum of the anterolateral abdominal wall can be detected by pressing the approximated fingers of one hand deep into a region of the patient's anterolateral abdominal wall and then suddenly withdrawing the fingers. The wave of sudden rebound movement which spreads throughout the abdominal walls elicits tenderness in those wall regions lined by inflamed parietal peritoneum. It is preferable to apply the finger pressure at a site distant from the suspected sites of inflammation. The examiner must use discretion in applying appropriate pressure during testing for rebound tenderness, as the pain can be quite severe. When an inflamed appendix lies in position B or D, its inflammatory process can readily extend posteriorly to the iliacus muscle in the right iliac fossa. Right iliopsoas inflammation is indicated if there is pain upon active flexion of the right thigh against resistance (which involves isometric contraction of the right iliopsoas muscle) or pain upon passive hyperextension of the right thigh (which involves stretching of the right iliopsoas muscle). When an inflamed appendix lies in position E, its inflammatory process can readily extend laterally to the obturator internus muscle in the right lateral pelvic wall or medially to the right side of the rectum. Right obturator internus inflammation is indicated if there is pain upon passive external and internal rotation of the right thigh, and right-sided rectal inflammation is indicated if there is right-sided rectal tenderness upon rectal exam.

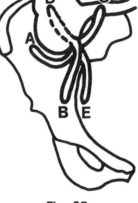

Fig. 25

Large Intestine

Items 345-347

The following items concern visceral pain associated with disease of the large intestine and the blood supply and parasympathetic innervation of the large intestine.

345. Outpouchings of the large intestine, called diverticula, sometimes develop in elderly individuals. These diverticula can become infected, and their inflammation can produce visceral pain. If a diverticulum of the sigmoid colon became inflamed, in which abdominal region will the individual most likely feel the visceral pain?

    (A) Left inguinal region
    (B) Left lumbar region
    (C) Epigastric region
    (D) Umbilical region
    (E) Hypogastric region

346. Where along the course of the large intestine do colic branches of the superior mesenteric artery anastomose with colic branches of the inferior mesenteric artery?

    (A) Near the border between the cecum and ascending colon
    (B) Near the hepatic flexure
    (C) Near the splenic flexure
    (D) Near the border between the descending colon and sigmoid colon
    (E) Near the border between the sigmoid colon and rectum

347. The vagus nerves provide preganglionic parasympathetic innervation to the large intestine as far distally as the region *as far as splenic flexure (Between middle third of Transverse colon + left third of Transverse colon)*

    (A) near the border between the cecum and ascending colon
    (B) near the hepatic flexure
    (C) near the splenic flexure
    (D) near the border between the descending colon and sigmoid colon
    (E) near the border between the sigmoid colon and rectum

pelvic splanchnic Nerve rest of large Intestine

ANSWERS AND TUTORIAL FOR ITEMS 345-347

The answers are: **345-E; 346-C; 347-C.** Disease or injury of the sigmoid colon and other abdominal viscera supplied by the inferior mesenteric artery produces visceral pain that is most commonly localized to the hypogastric region. The terminal branches of the ileocolic (IC), right colic (RC) and middle colic (MC) branches of the superior mesenteric artery extend distally along the large intestine as far the border region between the middle and left thirds of the transverse colon (Fig. 26). The terminal branches of the left colic (LC) and sigmoid (S) branches of the inferior mesenteric artery extend along the distal third of the transverse colon and the entire length of the descending and sigmoid colons. The anastomoses among the terminal branches of the colic branches of the superior and inferior mesenteric arteries form, in effect, an artery that runs along the medial margins of the ascending and descending colons and the mesenteric margins of the transverse and sigmoid colons. This artery is called the marginal artery of Drummond. The vagus nerves provide preganglionic parasympathetic innervation along the large intestine distally to the border between the middle and left thirds of the transverse colon. The pelvic splanchnic nerves provide preganglionic parasympathetic innervation to the remaining distal segments of the large intestine.

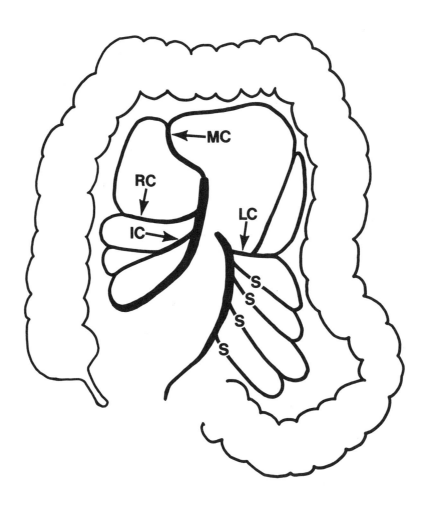

Fig. 26

A 31 year-old man complains of continuous excruciating abdominal pain that began 2 hours ago. A tentative diagnosis of ureteral obstruction is made on the basis of the history and physical exam. An intravenous urogram showing obstruction of the right ureter by a radiodense calculus at the ureteropelvic junction confirms the diagnosis.

348. The pain of ureteral colic may involve all of the following regions EXCEPT:

   (A) ipsilateral shoulder
   (B) ipsilateral costovertebral region
   (C) ipsilateral lumbar, or flank, region
   (D) ipsilateral inguinal region
   (E) ipsilateral scrotum or vulva

349. Sensory fibers from the upper part of the ureter enter the spinal cord at spinal cord segment levels

   (A) T1, T2, T3, T4 and T5
   (B) T5, T6, T7, T8 and T9
   (C) T8, T9, T10, T11 and T12
   (D) T10, T11, T12, L1 and L2
   (E) L1, L2, S2, S3 and S4

350. An individual suffering from acute pyelonephritis commonly exhibits tenderness upon gentle fist percussion of the back region overlying the inflamed kidney. The gentle fist percussion is applied immediately lateral to the vertebral column at the level of the

   (A) 6th thoracic vertebra
   (B) 8th thoracic vertebra
   (C) 10th thoracic vertebra
   (D) 12th thoracic vertebra
   (E) 4th lumbar vertebra

351. The distance between the superior and inferior poles of an adult kidney averages ___ times the thickness of the body of the 2nd lumbar vertebra.

   (A) 1.5
   (B) 2.6
   (C) 3.7
   (D) 4.8
   (E) 5.9

352. The adrenal glands lie atop the superior poles of the kidneys. All of the following statements concerning the blood supply and venous drainage of the adrenal glands are correct **EXCEPT:**

(A) Each adrenal gland is supplied by direct branches of the inferior phrenic artery.
(B) Each adrenal gland is supplied by direct branches of the abdominal aorta.
(C) Each adrenal gland is supplied by direct branches of the renal artery.
(D) The left adrenal vein commonly ends by union with the inferior vena cava.
(E) The right adrenal vein commonly ends by union with the inferior vena cava.

In the following items, match each macroscopic part of the kidney with its image in the labeled drawing below. The drawing shows a coronally sectioned view of the kidney.

353. Renal pyramid

354. Renal column

355. Minor calyx

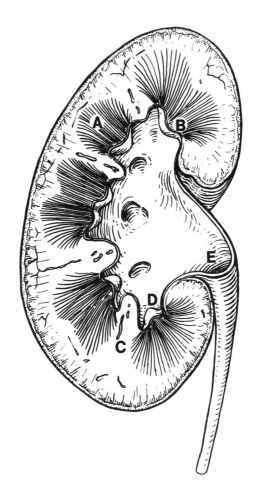

ANSWERS AND TUTORIAL FOR ITEMS 348-355

The answers are: **348-A 349-D; 350-D; 351-C; 352-D; 353-A; 354-C; 355-D.** Intravenous urograms permit evaluation of kidney position, size and parenchymal thickness; the shape of the renal papillae and calyces; and the presence of filling defects in the calyces and/or ureter. Disease or injury of the upper part of the ureter produces visceral pain that is frequently localized to the costovertebral region (the back region immediately inferior to the posterior part of the 12th rib and immediately lateral to the bodies of the 12th thoracic and 1st lumbar vertebrae). Disease or injury of the upper part of the ureter may refer pain to parts of the T10-L2 dermatomes, which include the flank and inguinal regions of the abdomen and the scrotum or vulva. Disease or injury of the kidney produces visceral pain that is most commonly localized to the costovertebral region. Deep tenderness upon gentle fist percussion to the costovertebral region (CVR) suggests kidney disease or injury (Fig. 27).

The envelope of perirenal fat occasionally permits visualization of the superior and inferior poles of a kidney in an abdominal plain film. Comparison of the length of the kidney relative to the thickness of the body of the 2nd lumbar vertebra can be used to assess if the kidney is within normal size limits.

The left adrenal vein commonly ends by union with the left renal vein. In the drawing of the coronally sectioned view of the kidney, the part labeled B is a renal papilla and the part labeled E is the pelvis of the ureter.

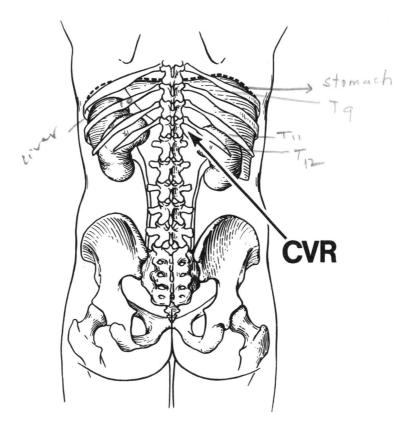

Fig. 27

The following items address information relevant to disease and disorders of pelvic and perineal viscera.

356. Which of the following structures most strongly supports the uterus in the pelvis?

   (A) Broad ligament of the uterus
   (B) Levator ani
   (C) Transverse cervical ligament
   (D) Round ligament of the uterus
   (E) Uterosacral ligament

357. Most of the lymph collected from the fundus of the uterus drains into the

   (A) aortic nodes
   (B) external iliac nodes
   (C) internal iliac nodes
   (D) horizontal group of superficial inguinal nodes
   (E) vertical group of superficial inguinal nodes

358. Most of the lymph collected from the lower half of the anal canal drains into the

   (A) aortic nodes
   (B) external iliac nodes
   (C) internal iliac nodes
   (D) horizontal group of superficial inguinal nodes
   (E) vertical group of superficial inguinal nodes

359. All of the following statements concerning digital examination of the anal canal and rectum are correct EXCEPT:

   (A) The intermuscular groove marks the lower border of the internal anal sphincter.
   (B) The anorectal ring marks the upper border of the external anal sphincter.
   (C) The prostate can be palpated through the anterior wall of the rectum.
   (D) The cervix of the uterus can generally be palpated through the anterior wall of the rectum.
   (E) The fundus of an anteverted and anteflexed uterus can generally be palpated through the anterior wall of the rectum.

360. All of the following statements concerning penile erection and ejaculation are correct **EXCEPT:**

    (A) The terminal branches of the internal pudendal arteries provide the main arterial supply of the erectile tissues of the penis.

    (B) Increased sympathetic activity during sexual arousal increases blood flow to the erectile tissues of the penis.

    (C) Penile erection occurs as a consequence of the engorgement of the cavernous sinuses in the corpora cavernosa of the penis.

    (D) Secretions from the seminal vesicles and prostate gland contribute to the composition of semen during emission.

    (E) Ejaculation of semen occurs as a consequence of the rhythmic contractions of the bulbospongiosus muscles.

361. All of the following statements concerning the bladder are correct **EXCEPT:**

    (A) Disease or injury of the bladder produces visceral pain that is most commonly localized to the suprapubic region.

    (B) Disease or injury of the bladder may refer pain to parts of the L1, L2, S2, S3 and S4 dermatomes.

    (C) The act of micturition requires voluntary relaxation of the sphincter urethrae.

    (D) Contraction of the anterolateral abdominal wall musculature and the diaphragm can aid micturition.

    (E) The act of micturition requires voluntary contraction of the levator ani.

ANSWERS AND TUTORIAL FOR ITEMS 356-361

The answers are: **356-B; 357-A; 358-D; 359-E; 360-B; 361-E.** The principal supports of the uterus are the paired levator ani muscles. The cardinal (transverse cervical) and the uterosacral ligaments, which are paired condensations of endopelvic fascia attached to the cervix of the uterus and vault of the vagina, provide secondary support. The cardinal ligaments extend from the lateral pelvic walls to the cervix and vagina along the lowest margin of the extraperitoneal space within the broad ligaments of the uterus. The cardinal ligaments help stabilize the midline position of the cervix and the vault of the vagina. The uterosacral ligaments arise from the lower end of the sacrum and extend anteriorly around the sides of the rectum to attach to the cervix and vagina. The uterosacral ligaments securely tether the cervix to the sacrum, and help stabilize the angle between the longitudinal axes of the vagina (V) and the cervix of the uterus (CU) (Fig. 28).

Lymph collected from the lower third of the vagina drains into the horizontal group of superficial inguinal nodes. Lymph collected from the middle third of the vagina drains into internal iliac nodes. Lymph collected from the upper third of the vagina drains into external and internal iliac nodes. Most of the lymph collected from the cervix and body of the uterus drains into internal iliac nodes. Lymph collected from the fundus of the uterus, the uterine tubes and the ovaries drains into aortic nodes at the level of the body of the 1st lumbar vertebra. Lymph collected from the upper half of the anal canal and the lower half of the rectum drains into internal iliac nodes. Lymph collected from the upper half of the rectum drains into the inferior mesenteric nodes.

When the urinary bladder (B) in a female is empty, it is common for the uterus to be both anteverted and anteflexed (Fig. 28). An anteverted uterus has its longitudinal axis bent forward (generally at an approximately 90 degree angle) relative to the longitudinal axis of the vagina (V). An anteflexed uterus has its body (BU) bent forward relative to the cervix (CU). The cervix of the uterus can generally be palpated through the anterior wall of the rectum (R). The fundus (FU) of an anteverted and anteflexed uterus may be palpated during bimanual examination of the pelvis. In a male, the prostate (P) can be palpated through the anterior wall of the rectum (R) (Fig. 29). Increased parasympathetic activity during sexual arousal increases blood flow to the erectile tissues of the penis. The act of micturition requires voluntary relaxation of the paired levator ani muscles and the sphincter urethrae. The relaxation of the levator ani not only pulls the neck of the bladder (NB) downward (and thus decreasing the resistance in the neck), but also reflexively stimulates contraction of the detrusor (Fig. 29). The detrusor is the smooth muscle tissue in the bladder wall. It is innervated by parasympathetic fibers provided via the pelvic splanchnic nerves.

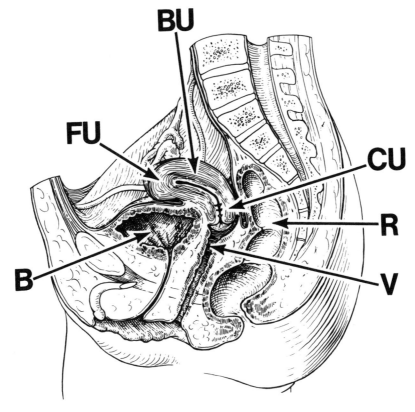

Fig. 28

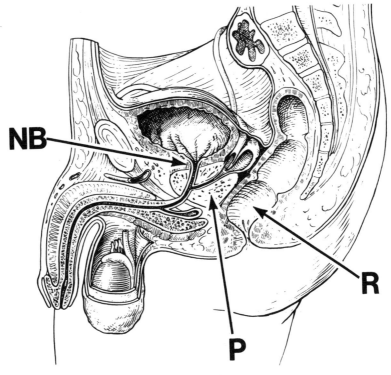

Fig. 29

The two CT scans below are adjacent 10 mm-thick abdominal sections. Segments of the upper gastrointestinal tract are partially filled with contrast material. In the following items, match each structure with its image in the labeled CT scan.

362. Inferior vena cava

363. Liver

364. Abdominal aorta

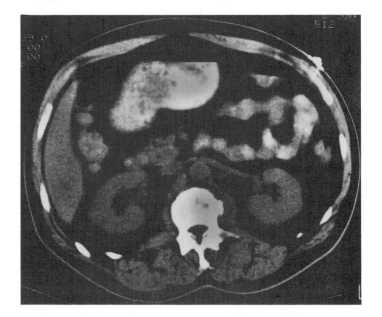

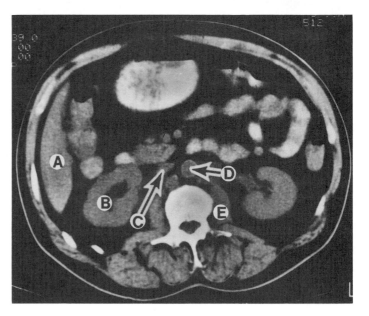

ANSWERS FOR ITEMS 362-364

The answers are: **362-C; 363-A; 364-D.** The structure labeled B is the right kidney, and the structure labeled E is the left psoas major muscle.

Items 365-366

A 23 year-old man hit his head on the ground upon falling from a bicycle. His cycling helmet protected him against skull fractures but did not prevent him from becoming permanently anosmic as a result of tearing of the first-order nerve fibers of both olfactory nerves as they extend from the nasal cavities to the cranial cavity.

365. Which part of the ethmoid bone transmits the nerve fibers of the olfactory nerve from the nasal cavity to the cranial cavity?

    (A) Perpendicular plate
    (B) Cribriform plate
    (C) Superior concha
    (D) Middle concha
    (E) Crista galli

366. The ethmoid bone contributes to the bony foundation of all of the following **EXCEPT**:

    (A) roof of the nasal cavity
    (B) lateral wall of the nasal cavity
    (C) nasal septum
    (D) floor of the nasal cavity
    (E) medial wall of the orbital cavity

ANSWERS AND TUTORIAL FOR ITEMS 365-366

The answers are: **365-B; 366-D.** The olfactory nerve (cranial nerve I) arises from first-order sensory neurons which lie within the mucosal lining of the roof of the nasal cavity. The afferent fibers of these first-order neurons extend from the nasal cavity into the cranial cavity by passing through perforations in the cribriform plate of the ethmoid bone. Upon entering the cranial cavity, these fibers extend up into the olfactory bulb to synapse with second-order neurons. Afferent fibers of the second-order neurons extend posteriorly through the olfactory tract to enter the brain. Blows to the head which displace the brain posteriorly can produce tears in the afferent fibers of the first-order neurons of the olfactory nerve near the sites where the fibers emerge from the perforations in the cribriform plate of the ethmoid bone.

    The cribriform plate of the ethmoid bone forms the bony foundation of the roof of the nasal cavity. The perpendicular plate of the ethmoid bone forms part of the bony foundation of the nasal septum. The superior and middle conchae form part of the bony foundation of the lateral wall of the nasal cavity. The orbital plate of the ethmoid bone forms part of the bony foundation of the medial wall of the orbital cavity. The palatine process of the maxillary bone and the horizontal plate of the palatine bone form the bony foundation of the floor of the nasal cavity.

Items 367-369

Invasive tumors of the sphenoid bone may press upon and impair the nerves that extend through the superior orbital fissure.

367. All of the following nerves extend through the superior orbital fissure EXCEPT:

(A) ophthalmic division of the trigeminal nerve
(B) oculomotor nerve
(C) optic nerve
(D) trochlear nerve
(E) abducent nerve

368. All of the following alterations could occur as a result of injury to the nerves that extend through the superior orbital fissure EXCEPT:

(A) The ability to move the eyeball from the primary position to a position in which the cornea faces directly laterally could be weakened or lost.
(B) The ability to move the eyeball from the primary position to a position in which the cornea faces directly medially could be weakened or lost.
(C) The ability to forcibly close the eyelids could be weakened or lost.
(D) The ability to raise the upper eyelid could be weakened or lost.
(E) The ability to decrease the size of the pupil could be weakened or lost.

369. All of the following alterations could occur as a result of injury to the nerves that extend through the superior orbital fissure EXCEPT:

(A) The ability to sense a touch to the forehead could be diminished or lost.
(B) The ability to sense a touch to the tip of the nose could be diminished or lost.
(C) The ability to sense a touch to the cornea could be diminished or lost.
(D) The ability to sense a touch to the upper eyelid could be diminished or lost.
(E) The ability to sense a touch to the lower eyelid could be diminished or lost.

166

ANSWERS AND TUTORIAL FOR ITEMS 367-369

The answers are: **367-C; 368-C; 369-E.** The superior orbital fissure is a slit between the greater and lesser wings of the sphenoid bone. It transmits the oculomotor nerve, trochlear nerve, ophthalmic division of the trigeminal nerve and abducent nerve between the cranial and orbital cavities.

The optic nerve (cranial nerve II) extends from the orbital to the cranial cavity by passing through the optic canal, a passageway that traverses the body and lesser wing of the sphenoid bone.

The oculomotor nerve (cranial nerve III) innervates four of the extraocular muscles of the eyeball (medial rectus, superior rectus, inferior rectus and inferior oblique) and levator palpebrae superioris. Medial rectus moves the eyeball medially. Levator palpebrae superioris raises the upper eyelid. The oculomotor nerve also transmits preganglionic parasympathetic fibers which synapse with postganglionic parasympathetic neurons in the ciliary ganglion. The ciliary ganglion lies in the orbital cavity posterior to the eyeball. The postganglionic parasympathetic fibers which emanate from the ciliary ganglion innervate two intraocular muscles: sphincter pupillae (which acts to decrease the size of the pupil) and ciliaris (which acts to focus the lens on near objects).

The trochlear nerve (cranial nerve IV) innervates the extraocular muscle superior oblique. The ophthalmic division of the trigeminal nerve (cranial nerve V) provides sensory innervation for (a) the mucous membrane lining the anterior third of the nasal cavity, frontal sinus, sphenoid sinus and part of the ethmoid air cells, (b) the bulbar conjunctiva of the eyeball and the palpebral conjunctiva of the upper eyelid and (c) the skin of the forehead, the upper eyelid and the bridge and tip of the nose. The maxillary division of the trigeminal nerve provides sensory innervation for the skin of the lower eyelid.

The abducent nerve (cranial nerve VI) innervates lateral rectus. When acting alone, lateral rectus moves the eyeball laterally. The facial nerve (cranial nerve VII) innervates orbicularis oculi (OO), the muscle of facial expression responsible for forcibly closing the eyelids (Fig. 30).

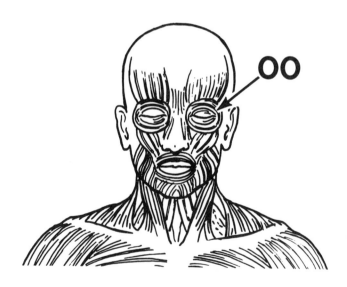

Fig. 30

Items 370-372

A 38 year-old woman with septic thrombosis of the cavernous sinus complains of double vision whenever she looks downward. The two images are farthest apart when she looks downward and to the left. The physical exam shows that the patient can move her right eyeball directly to the left. However, when the cornea of her right eyeball faces directly medially, the patient is unable to move her right eyeball so that the cornea also looks downward.

370. Which muscle of her right eye is not acting effectively?

    (A) Superior rectus
    (B) Inferior rectus
    (C) Lateral rectus
    (D) Inferior oblique
    (E) Superior oblique

371. All of the following structures extend through the interior of the cavernous sinus or its lateral wall **EXCEPT:**

    (A) postganglionic sympathetic fibers that innervate dilator pupillae
    (B) internal carotid artery
    (C) trochlear nerve
    (D) maxillary division of the trigeminal nerve
    (E) mandibular division of the trigeminal nerve

372. All of the following statements concerning the cavernous sinus are correct **EXCEPT:**

    (A) The cavernous sinus lies immediately lateral the pituitary gland.
    (B) The cavernous sinus lies immediately medial to the frontal lobe of the cerebrum.
    (C) The cavernous sinus can drain blood from the facial vein.
    (D) The superior ophthalmic vein is a tributary of the cavernous sinus.
    (E) The internal jugular vein can drain blood from the cavernous sinus.

ANSWERS AND TUTORIAL FOR ITEMS 370-372

The answers are: **370-E; 371-E; 372-B.** The cavernous sinus lies medial to the temporal lobe of the cerebral hemisphere. Septic thrombosis of the cavernous sinus may inflame and injure the nerves that extend through the interior of the sinus or its lateral wall. The lateral wall of the cavernous sinus transmits segments of the oculomotor nerve, the ophthalmic and maxillary divisions of the trigeminal nerve and the trochlear nerve. The interior of the cavernous sinus is traversed by segments of the abducent nerve and the internal carotid artery (and accompanying postganglionic sympathetic fibers, some of which innervate dilator pupillae).

The medial and lateral rectus muscles move the eyeball from its primary position to orientations in which the cornea faces, respectively, directly medially and laterally. The activities of the medial and lateral rectus muscles are tested during a physical exam by asking the patient to look in directions that directly match the actions of these muscles. For example, to test the activity of the medial rectus, the patient is asked to try to look at the nose. In this case involving a patient with diplopia, the medial rectus of the right eye functions normally.

The superior and inferior rectus muscles move the eyeball from its primary position to orientations in which the cornea faces, respectively, superomedially and inferomedially. The superior and inferior oblique muscles move the eyeball from its primary position to orientations in which the cornea faces, respectively, inferolaterally and superolaterally. However, in testing the activities of the superior and inferior recti and the superior and inferior obliques, the patient is asked to look in directions that do not correspond to the actions of these muscles. With the cornea facing directly laterally (that is, with the patient looking directly to the side of the head), the activities of the superior and inferior rectus muscles are tested by asking the patient to look, respectively, upward and downward. With the cornea facing directly medially (that is, with the patient looking directly at the nose), the activities of the superior and inferior obliques are tested by asking the patient to look, respectively, downward and upward. Fig. 31 is a drawing of the superior view of the relationship of the superior oblique muscle (SO) to the right eyeball. Notice that if the cornea (C) faces directly medially, superior oblique is the only extraocular muscle whose actions can move the eyeball so that the cornea also faces inferiorly.

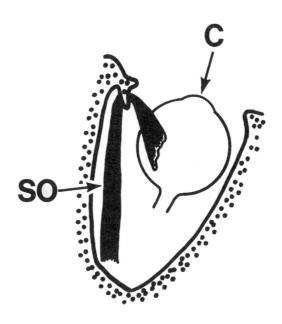

Fig. 31

The following items pertain to tests that can be performed to examine cranial nerve function.

373. The jaw jerk is a deep tendon reflex in which tapping a patient's lower jaw causes a reflexive raising of the lower jaw. The jaw jerk tests sensory fibers of the

(A) mandibular division of the trigeminal nerve
(B) facial nerve
(C) glossopharyngeal nerve
(D) vagus nerve and cranial root of the accessory nerve
(E) hypoglossal nerve

374. The jaw jerk tests motor fibers of the

(A) mandibular division of the trigeminal nerve
(B) facial nerve
(C) glossopharyngeal nerve
(D) vagus nerve and cranial root of the accessory nerve
(E) hypoglossal nerve

375. An examiner palpates a patient's masseter muscles as the patient is asked to clench the teeth. This procedure tests motor fibers of the

(A) mandibular division of the trigeminal nerve
(B) facial nerve
(C) glossopharyngeal nerve
(D) vagus nerve and cranial root of the accessory nerve
(E) hypoglossal nerve

376. During mirror laryngoscopy, an examiner notes paralysis of the right vocal fold (the right vocal fold is not abducted when the patient is requested to say "e-e-e" in a high-pitched voice). The observation indicates paralysis of muscles innervated by the

(A) mandibular division of the trigeminal nerve
(B) facial nerve
(C) glossopharyngeal nerve
(D) vagus nerve and cranial root of the accessory nerve
(E) hypoglossal nerve

ANSWERS AND TUTORIAL FOR ITEMS 373-376

The answers are: **373-A; 374-A; 375-A; 376-D.** The stretch receptors and muscle fibers of the muscles of mastication (temporalis, masseter, lateral pterygoid and medial pterygoid) are innervated by the mandibular division of the trigeminal nerve. Masseter and medial pterygoid raise the lower jaw at the temporomandibular joint (TMJ). The tone of the masseters increases when an individual clenches the teeth. Temporalis raises and retracts the lower jaw at the TMJ. Lateral pterygoid lowers and protracts the lower jaw.

The intrinsic muscles of the larynx (which are the muscles that abduct, adduct, tense and relax the vocal folds) are innervated by branches of the vagus nerves. Almost all of the motor nerve fibers to the intrinsic muscles of the larynx are from the cranial root of the accessory nerve (which joins the vagus nerve in the uppermost part of the neck). With the exception of the cricothyroids, all the intrinsic muscles of the larynx are innervated by the recurrent laryngeal nerves. The cricothyroids are innervated by the external laryngeal nerves. The muscles chiefly responsible for abducting the vocal folds during phonation are the posterior cricoarytenoids.

Items 377-383

The following items pertain to tests that can be performed to examine cranial nerve function.

377. An examiner asks a patient to stick out the tongue and notes a left deviation. If the deviation is a result of disease or injury to the motor fibers in or more nerves, then those motor fibers are in the

    (A) mandibular division of the left trigeminal nerve
    (B) left vagus nerve and cranial root of the left accessory nerve
    (C) right vagus nerve and cranial root of the right accessory nerve
    (D) left hypoglossal nerve
    (E) right hypoglossal nerve

378. An examiner asks a patient to stick out the tongue and say "ah" and notes left deviation of the uvula. If the deviation is a result of disease or injury to the motor fibers in or more nerves, then those motor fibers are in the

    (A) left glossopharyngeal nerve
    (B) left vagus nerve and cranial root of the left accessory nerve
    (C) right vagus nerve and cranial root of the right accessory nerve
    (D) left hypoglossal nerve
    (E) right hypoglossal nerve

379. An examiner asks a patient to smile and notes that the left corner of the patient's mouth does not move upward. If this abnormality is a result of disease or injury to the motor fibers in or more nerves, then those motor fibers are in the

    (A) mandibular division of the left trigeminal nerve
    (B) left facial nerve
    (C) left glossopharyngeal nerve
    (D) left vagus nerve and cranial root of the left accessory nerve
    (E) left hypoglossal nerve

172

380. An examiner massages a patient's neck in the region overlying the carotid pulse and notes a reflexive decrease in the rate of the patient's heartbeat. This reflex is mediated by preganglionic parasympathetic fibers of the

(A) facial nerve
(B) glossopharyngeal nerve
(C) vagus nerve
(D) cranial root of the accessory nerve
(E) hypoglossal nerve

381. The nerve or nerves that provide taste fibers for sugar on the tip of the tongue is/are the

(A) mandibular division of the trigeminal nerve
(B) facial nerve
(C) glossopharyngeal nerve
(D) vagus nerve and cranial root of the accessory nerve
(E) hypoglossal nerve

382. The nerve or nerves that provide sensation for light touch on the tip of the tongue is/are the

(A) mandibular division of the trigeminal nerve
(B) facial nerve
(C) glossopharyngeal nerve
(D) vagus nerve and cranial root of the accessory nerve
(E) hypoglossal nerve

383. The muscle or muscles responsible for sticking out the tongue is/are the

(A) intrinsic muscles of the tongue
(B) palatoglossus
(C) genioglossus
(D) styloglossus
(E) hyoglossus

ANSWERS AND TUTORIAL FOR ITEMS 377-383

The answers are: **377-D; 378-C; 379-B; 380-C; 381-B; 382-A; 383-C.** Genioglossus, hyoglossus, styloglossus and all the intrinsic muscles of the tongue are innervated by the hypoglossal nerve. Genioglossus is the only extrinsic muscle of the tongue which can protrude the tongue. When genioglossus acts to protrude the tip of the tongue, it also deviates the tip of the tongue to the contralateral side. Consequently, denervation of the left genioglossus will cause left lingual deviation, since the left genioglossus cannot oppose the leftward thrust of the right genioglossus.

Musculus uvula and levator veli palatini raise and pull the uvula to the ipsilateral side. Denervation the left musculus uvula and levator veli palatini will cause the uvula to deviate to the right when the individual says "ah," since the left musculus uvula and levator veli palatini cannot oppose the rightward pull of the right musculus uvula and levator veli palatini. Almost all of the motor nerve fibers to musculus uvula and levator veli palatini are from the cranial root of the accessory nerve (which joins the vagus nerve in the uppermost part of the neck). The muscles of facial expression are responsible for the appearance of a smile, and are innervated by the terminal branches of the facial nerve.

The origin of the internal carotid artery commonly exhibits a dilatation called the carotid sinus (the carotid sinus may also be located at the termination of the common carotid artery). Each carotid sinus bears many baroreceptors innervated chiefly by the glossopharyngeal nerve. When these fibers are subjected to a sudden change in blood pressure, their response initiates an autonomic reflex which restores the blood pressure back to normal levels. The restoration occurs via regulation of arteriolar constriction and heart rate. A sudden increase in blood pressure stimulates the parasympathetic innervation of the cardiac plexuses by the vagus nerves. The parasympathetic stimulation decreases the heart rate.

The chorda tympani branch of the facial nerve provides taste sensation for the anterior two-thirds of the tongue. The mandibular division of the trigeminal nerve provides sensory innervation for the anterior two-thirds of the tongue. The glossopharyngeal nerve provides both sensory innervation and taste sensation for the posterior third of the tongue.

174

The functions served by the facial nerve include the following: *7th CN*

[1]   Its greater petrosal branch provides preganglionic parasympathetic innervation for the lacrimal gland and the mucosal glands of the nasal cavity. The preganglionic parasympathetic fibers synapse with postganglionic parasympathetic neurons in the pterygopalatine ganglion.

[2]   It innervates stapedius, a muscle of the middle ear.

[3]   Its chorda tympani branch provides taste sensation for the anterior two-thirds of the tongue and preganglionic parasympathetic innervation for the submandibular and sublingual salivary glands. The preganglionic parasympathetic fibers synapse with postganglionic parasympathetic neurons in the submandibular ganglion.

[4]   Its terminal branches innervate stylohyoid, the posterior belly of digastric and all the muscles of facial expression.

The functions served by the glossopharyngeal nerve include the following: *9th CN*

[1]   Its lesser petrosal branch provides preganglionic parasympathetic innervation for the parotid salivary gland. The preganglionic parasympathetic fibers synapse with postganglionic parasympathetic neurons in the otic ganglion.

[2]   It innervates stylopharyngeus.

[3]   It provides sensory innervation for the posterior third of the tongue, the lower half of the nasopharynx and all of the oropharynx and laryngopharynx. The vagus nerve also provides sensory innervation for the oropharynx and laryngopharynx.

[4]   It provides taste sensation for the posterior third of the tongue.

[5]   It supplies the baroreceptors of the carotid sinus.

[6]   It supplies the chemoreceptors of the carotid body.

The two CT scans below are adjacent 10 mm-thick scans of an individual's head. Contrast material was administered intravenously prior to the scans. Match each item with its image in the labeled CT scan.

384. Mastoid air cells

385. Sphenoid sinuses

386. External auditory meatus

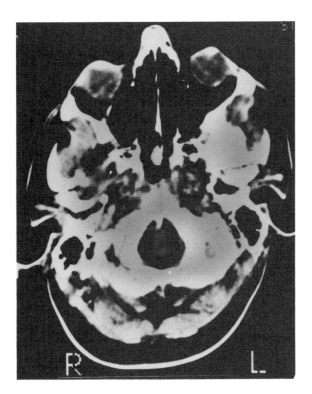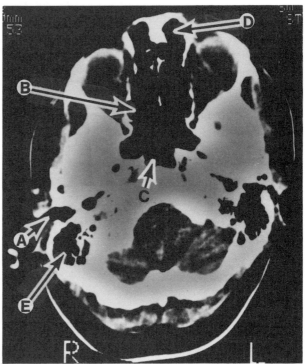

ANSWERS FOR ITEMS 384-386

The answers are: **384-E; 385-C; 386-A**. The space labeled B is the right ethmoid air cells, and the space labeled D is the left frontal sinus.

Chapter IV

NEUROANATOMY

<u>Items 387-391</u>

The diagram below represents a cross-section through the brainstem.

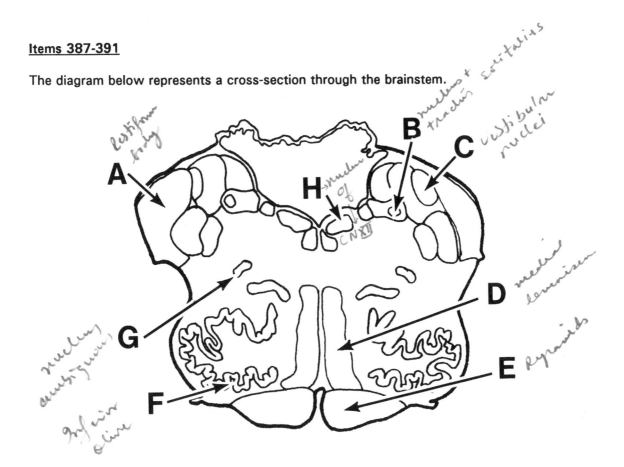

387. The cross-section represents what portion of the brainstem?

    (A) Rostral pons
    (B) Midbrain
    (C) Rostral medulla
    (D) Caudal medulla
    (E) Caudal pons

388. Which labeled structure carries proprioceptive information from the limbs?

    (A) A
    (B) B
    (C) C
    (D) D
    (E) E

177

389. The origin of most of the axons found in structure E is the

    (A) cerebellum
    (B) motor and premotor cortex
    (C) spinal cord
    (D) caudate nucleus
    (E) red nucleus

390. Which structure provides motor innervation to the muscles of the soft palate and pharynx?

    (A) D
    (B) E
    (C) F
    (D) G
    (E) H

391. The portion of the ventricular system found at this level of the brainstem is the

    (A) open portion of the 4th ventricle
    (B) cerebral aqueduct
    (C) 3rd ventricle
    (D) central canal
    (E) lateral ventricle

ANSWERS AND TUTORIAL ON ITEMS 387-391

The answers are: **387-C; 388-D; 389-B; 390-D; 391-A.** The diagram is a cross-section through the rostral portion of the medulla. Structures found at this level include: A - restiform body; B - nucleus and tractus solitarius; C - vestibular nuclei; D - medial lemniscus; E - pyramids; F - inferior olive; G - nucleus ambiguus; and H - nucleus of cranial nerve XII (hypoglossal). The portion of the ventricular system seen at this level is the open portion of the 4th ventricle. The inferior olive (F) is probably the most distinctive structure seen in cross-section at this level of the brainstem. The pathway responsible for carrying proprioceptive information from the spinal cord is the medial lemniscus (D). The origin of most of the axons found in the pyramids (E) is from the motor and premotor cortices of the frontal lobes as well as from primary sensory cortex (postcentral gyrus). The nucleus ambiguus (G) provides axons that are carried by cranial nerve X (as well as cranial nerves IX and XI) for innervation of the muscles of the soft palate, pharynx and larynx.

178

Examination of a 35 year-old man reveals that he has a severe impairment of visual acuity, blurring of vision and diplopia. The patient has pronounced weakness of all four extremities, the deep tendon reflexes are exaggerated and he has bilateral Babinski signs. The patient complains of feelings of numbness and tingling in the extremities, trunk and face. He also describes a sensation of 'electricity' on passive or active flexion of the neck. Examination reveals impairment of vibratory, position and pain sensation in the extremities. The nurse states that the man has urinary incontinence. The patient's speech is slurred and somewhat slow and has a sing-song quality (scanning speech). On attempting to stand and walk, the patient shows tremors and incoordination of the muscles of the trunk and extremities and ataxia of gait. The patient states that his symptoms were not always as severe as they are now, but they tended to vary in nature and severity over a number of years.

392. Given these signs and symptoms, a likely diagnosis might be

    (A) multiple sclerosis
    (B) Huntington's chorea
    (C) tabes dorsalis
    (D) Friedreich's ataxia
    (E) Parkinson's disease

393. The patient's slurred speech and scanning speech are indicative of damage to the

    (A) anterior limb of the internal capsule
    (B) pyramids and pyramidal decussation
    (C) posterior columns and posterior column nuclei
    (D) caudate nucleus and putamen
    (E) cerebellum or its connections with the deep cerebellar nuclei

394. The patient's impairment of vibratory and position sense are indicative of damage to the

    (A) corticospinal tract
    (B) spinothalamic tract
    (C) pyramids
    (D) spinocerebellar tracts
    (E) posterior columns

395. The patient's exaggerated deep tendon reflexes and the bilateral Babinski signs are indicative of damage to the

    (A) spinocerebellar tracts
    (B) posterior columns
    (C) corticospinal tracts
    (D) spinothalamic tracts
    (E) lower motor neurons

two images of single object

396. The blurred vision and diplopia might indicate damage to the

    (A) corticospinal tracts
    (B) medial longitudinal fasciculus
    (C) medial lemniscus
    (D) superior colliculus
    (E) tractus solitarius

ANSWERS AND TUTORIAL ON ITEMS 392-396

The answers are: **392-A; 393-E; 394-E; 395-C; 396-B.** Multiple sclerosis is primarily a disease of the white matter and is classified as a demyelinating disease. The signs and symptoms of multiple sclerosis are so diverse that their enumeration would include all of the symptoms which can result from injury to any part of the neuraxis from the spinal roots to the cerebral cortex. Moreover, the signs and symptoms tend to vary in nature and severity with the passage of time. Retrobulbar neuritis is a very common manifestation of multiple sclerosis leading to a profound loss of visual acuity. Diplopia in multiple sclerosis may be caused by involvement of the medial longitudinal fasciculus resulting in an internuclear ophthalmoplegia. Involvement of the pyramidal tracts gives rise to spasticity, hyperreflexia and bilateral Babinski signs. Weakness of the extremities is the most common sign of the disease and may be manifested as monoplegia, hemiplegia, paraplegia or quadriplegia. The cerebellum or its connections with the brainstem are involved in the majority of cases giving rise to speech disturbances (slurred and scanning speech), ataxia of gait, tremors and incoordination of the muscles of the trunk and extremities. Urinary disturbances are also common, including incontinence. Damage to the posterior columns produces paresthesias including spontaneous feelings of numbness and tingling in the extremities, trunk or face.

Items 397-400

The neuroradiologist asks you to compare normal with abnormal arteriograms of the circle of Willis and cerebral arterial supply. You correctly note that the abnormal film shows a complete occlusion of the right internal carotid artery where it lies within the cavernous sinus.

397. One of the symptoms shown by the patient with the abnormal film is blindness in the right eye. This is due to compromise of which branch of the internal carotid artery?

    (A) lenticulostriate
    (B) ophthalmic
    (C) middle cerebral
    (D) anterior cerebral
    (E) posterior communicating

398. The branch of the internal carotid artery which provides most of the blood supply to the lateral aspect of the frontal and parietal lobes is the

    (A) anterior cerebral
    (B) middle cerebral
    (C) posterior cerebral
    (D) superior cerebellar
    (E) basilar

399. Occlusion of which artery would most severely compromise the blood supply to the visual cortex?

   (A) Posterior cerebral
   (B) Middle cerebral
   (C) Anterior cerebral
   (D) Basilar
   (E) Internal carotid

400. The oculomotor nerve passes between which two arteries at the base of the brain and may be affected by an aneurysm of either of these arteries?

   (A) Middle cerebral - anterior cerebral
   (B) Anterior cerebral - anterior communicating
   (C) Superior cerebellar - anterior inferior cerebellar
   (D) Posterior cerebral - superior cerebellar
   (E) Basilar - vertebral

ANSWERS AND TUTORIAL ON ITEMS 397-400

The answers are: **397-B; 398-B; 399-A; 400-D**. One of the first major branches of the internal carotid artery is the ophthalmic artery which in turn gives rise to the central retinal artery. Occlusion of either the ophthalmic or its branch, the central retinal artery, could result in blindness. The vessels which provide most of the blood supply to the lateral aspects of the frontal and parietal lobes of the cerebral hemispheres are the middle cerebral arteries. The posterior cerebral arteries provide most of the blood supply to the occipital lobes which are the location of the visual cortices. The oculomotor nerve (CN III) exits the brainstem between the posterior cerebral and the superior cerebellar arteries and thus an aneurysm of either artery could compromise the nerve.

Anterior Cerebral Artery →

Items 401-405

A 57 year-old man was admitted to the emergency room complaining of sharp pains in his lower extremities. He described the pains as being like 'jabs of lightning' or the 'stick of a sharp needle'. He stated that the pains would come and go over the past three years but they have become more frequent during the last few months. On testing it is noted that the man walks with an uncertain gait, shows an incoordination in the movements of the legs and a noticeable slapping of the feet. The patient shows Romberg's sign. Deep tendon reflexes at the ankle and knee are absent and the patient lacks vibratory and position sense in both lower extremities. Muscle strength appears to be normal. Cranial nerves appear intact but the pupils of both eyes are constricted, fail to respond to light but show normal reaction to accommodation-convergence. When queried about previous illness, the patient admits to having had syphilis when he was younger.

401. Given these symptoms and description, a likely diagnosis would be

   (A) syringomyelia
   (B) multiple sclerosis
   (C) amyotrophic lateral sclerosis
   (D) tabes dorsalis
   (E) Friedreich's ataxia

402. The pupillary responses shown by this patient might indicate a lesion in the

   (A) lateral geniculate nucleus
   (B) nucleus of Edinger-Westphal
   (C) pretectal region
   (D) cranial nerve III
   (E) Meyer's loop

403. The loss of vibratory and position sense in this patient are explainable by damage to the

   (A) ventral roots
   (B) posterior roots and posterior columns of the spinal cord
   (C) lateral white column of the spinal cord
   (D) anterior white commissure of the spinal cord
   (E) corticospinal tracts

404. A patient with Romberg's sign would show

   (A) a tendency to sway and fall when asked to stand with the feet together and the eyes closed
   (B) a dorsiflexion of the large toe and a fanning of the lateral four toes
   (C) inability to touch the tip of the nose with the index finger
   (D) inability to correctly determine a figure such as a numeral drawn on the skin by the examiner
   (E) loss of pain and temperature sensation from the affected part of the body

182

405. The most likely explanation for the ataxia in this patient is

    (A) loss of ventral horn motor neurons
    (B) damage to the cerebellum
    (C) loss of position sense in the affected extremities
    (D) loss of descending input from the cerebral cortex
    (E) loss of lateral horn motor neurons

ANSWERS AND TUTORIAL ON ITEMS 401-405

The answers are: **401-D; 402-C; 403-B; 404-A; 405-C.** The triad of symptoms - lancinating or lightning pains, ataxia, and dysuria together with the triad of signs - Argyll-Robertson pupils, absence of deep tendon reflexes and the loss of proprioception are characteristic of tabes dorsalis (progressive locomotor ataxia). Tabes is a late manifestation of neurosyphilis and is characterized pathologically by degenerative changes in the posterior roots, in the posterior funiculi of the spinal cord and in the brainstem. The destruction of the posterior roots explains the lancinating pains felt by these patients. Later, destruction of the posterior columns accounts for the loss of reflexes and the appearance of Romberg's sign (with eyes closed and feet together, the patient will sway and fall). Destruction of the posterior columns also accounts for the loss of vibratory and position sense and leads to ataxia in the patient. Involvement of the pretectal region of the midbrain accounts for the Argyll-Robertson pupil.

in the
cerebral or
medullar or
upper cervical
cord

↓
superior colliculus

Items 406-409

A 40 year-old woman complains that for several months she has had a weakness in both of her hands and has noted that when she touches a hot iron or places her hands in scalding hot water, she feels no pain. On testing it is found that she exhibits signs of wasting of the small muscles of both hands and there is a loss of pain and temperature sensation in the cervical and upper thoracic dermatomes in a shawl-like distribution, with prominent loss of pain and temperature sensation in the hands and forearms. Deep tendon reflexes are absent in the arms.

406. Which of the following diagrams of spinal cord cross-sections most accurately depicts the location and extent of the damage in this particular patient?

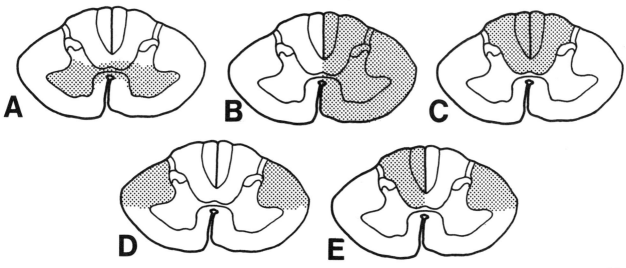

407. One likely diagnosis for this patient given the signs and symptoms is

   (A) poliomyelitis
   (B) Horner's syndrome
   (C) cerebrovascular accident
   (D) cervical syringomyelia
   (E) carotid aneurysm

408. Wasting of the small muscles of the hand is due to damage or destruction of the

   (A) corticospinal tracts
   (B) rubrospinal tracts
   (C) spinothalamic tracts
   (D) ventral horn motor neurons
   (E) posterior columns

409. The bilateral loss of pain and temperature sense is due to damage or destruction of the

   (A) anterior white commissure
   (B) spinothalamic tract
   (C) corticospinal tract
   (D) posterior columns
   (E) ventral horn motor neurons

ANSWERS AND TUTORIAL ON ITEMS 406-409

The answers are: **406-A; 407-D; 408-D; 409-A.** The patient shows the classical and most common form of syringomyelia. This is a disease of the spinal cord of unknown cause and is characterized pathologically by gliosis and cavitation of variable extent. This disease most often affects the cervical region of the cord (cross-section A) but may affect the lumbar cord and lower medulla as well. Patients experience loss of pain and temperature sensation due to destruction of the pain fibers crossing through the anterior white commissure. As the cavitation increases, the ventral horn motor neurons will be affected leading to atrophy of the denervated muscles. Atrophy of the small muscles of the hand localize the lesion in this patient to spinal segments C8 and T1. The loss of reflexes is probably due to the destruction of the posterior root collaterals to the ventral horn motor neurons and to loss of the motor neurons themselves.

Items 410-413

A 51 year-old women presents with a partial weakness of the muscles of the right side of her face, conjunctivitis and a corneal ulcer on the same side. On testing, it is found that she has cutaneous hypesthesia of the right side of her face and loss of the corneal reflex on the right side. Further testing shows that she also has a complete loss of response to caloric stimulation on the right and a horizontal nystagmus. Auditory testing reveals she has nerve deafness in the right ear. Following lumbar puncture, her CSF shows an increased protein content and skull X-rays reveal an enlargement of the internal acoustic meatus. A CT scan shows displacement and rotation of the fourth ventricle and changes in the cerebellopontine angle.

410. Given these signs and symptoms, a likely diagnosis would be

    (A) multiple sclerosis
    (B) Friedreich's ataxia
    (C) Korsakoff's syndrome
    (D) tic douloureux
    (E) acoustic neuroma

411. The loss of the caloric response and nystagmus indicate involvement of the

    (A) superior colliculus
    (B) medial geniculate body
    (C) vestibular portion of cranial nerve VIII
    (D) trigeminal nerve
    (E) facial nerve

412. The absence of the corneal reflex and hypesthesias of the face indicate involvement of the

    (A) trigeminal nerve
    (B) facial nerve
    (C) vagus nerve
    (D) glossopharyngeal nerve
    (E) chorda tympani

413. Another deficit that this patient may show with careful testing is loss of taste on the anterior 2/3 of the right side of the tongue. This is explainable by damage to cranial nerve

    (A) VIII
    (B) VII
    (C) IX
    (D) X
    (E) XII

ANSWERS AND TUTORIAL ON ITEMS 410-413

The answers are: **410-E; 411-C; 412-A; 413-B.** Tumors of cranial nerve VIII account for about 10 percent of all intracranial tumors and probably arise in the Schwann cell sheath of the nerve. These tumors grow slowly and symptoms are usually present for many months or years before the diagnosis is made. Damage to the auditory portion of cranial nerve VIII results in loss of hearing from the affected ear. The tumor frequently arises along the nerve in the internal acoustic meatus which accounts for the enlargement of the meatus seen in an X-ray. As the tumor enlarges, it compresses surrounding structures and affects nearby cranial nerves. Compression of the brainstem results in the distortion of the 4th ventricle and cerebellopontine angle as seen in the CT scan. The loss of the response to caloric stimulation and the horizontal nystagmus are due to destruction of the vestibular portion of nerve VIII. The trigeminal nerve is affected in 50% of the cases and results in the loss of sensation and loss of the corneal reflex on the same side as the tumor. Cranial nerve VII (facial) passes through the internal acoustic meatus along with cranial nerve VIII, and thus is frequently involved. Damage to cranial nerve VII results in paralysis or weakness of the facial muscles, loss of tearing from the ipsilateral eye and diminished taste sensation from the ipsilateral side of the tongue (anterior 2/3). The conjunctivitis and corneal ulcers are due to drying of the eye and the inability to close the eyelids (paralysis of the orbicularis oculi muscle).

Items 414-417

A newborn infant, on gross examination, has an obvious spina bifida in the lumbar region with meningomyelocele and hydrocephalus. X-ray (ventriculography) and CT examination reveal enlargement of the lateral ventricles, and shows that the 4th ventricle lies at the level of the foramen magnum.

414. Spina bifida is defined as

   (A) a failure of formation of the meninges
   (B) a failure in the closure of the spinal column due to a defect in the development of vertebrae
   (C) a failure of the proper formation of spinal nerve roots
   (D) an enlargement of the ventricular system of the brain
   (E) a failure of neural tube morphogenesis

415. The position of the 4th ventricle at the level of the foramen magnum indicates

   (A) the infant has a perfectly normal ventricular system
   (B) herniation of the brainstem and cerebellum through the foramen magnum (Arnold-Chiari malformation)
   (C) there is blockage of the ventricular system at the interventricular foramina
   (D) there is a defect in the occipital bone of the skull
   (E) there is a defect in the articulation between the atlas and the skull

416. The term meningomyelocele indicates that

   (A) the spinal cord and nerve roots are intact and only a herniated meningeal sac is present
   (B) the lumbar vertebrae lack a neural arch but the spinal cord and spinal nerve roots are intact and no meningeal sac is present
   (C) a portion of the spinal cord has been distorted, the spinal nerve roots are stretched and the cord is in a superficial position in the herniated meningeal sac
   (D) the central canal of the spinal cord has become enlarged
   (E) the ventricular system of the brain has become enlarged

417. This child would likely exhibit all of the following **EXCEPT**:

    (A) normal cutaneous and proprioceptive sensation in the lower extremities
    (B) loss of deep tendon reflexes in the lower extremities
    (C) urinary incontinence
    (D) weakness or paralysis of the leg muscles
    (E) atrophy of leg muscles

ANSWERS AND TUTORIAL ON ITEMS 414-417

The answers are: **414-B; 415-B; 416-C; 417-A.** Spina bifida is a failure in the closure of the spinal column due to a defect in the development of vertebrae. It may be classified as spina bifida occulta, spina bifida with meningocele or spina bifida with meningomyelocele depending on the severity of the involvement of the meninges, spinal cord and nerve roots. Spina bifida occulta indicates a simple defect only in the closure of the vertebrae. Meningocele indicates the presence of a sac-like protrusion of skin and meninges while meningomyelocele indicates the presence of a sac containing spinal cord and nerve roots. With the meningomyelocele defect in the lumbar region as this child shows, the signs and symptoms include: weakness and atrophy of leg muscles, disturbances of gait, urinary incontinence, loss of deep tendon reflexes of the legs and impairment of cutaneous and proprioceptive sensations in the lower extremities. The Arnold-Chiari malformation is a herniation of the brainstem and cerebellum through the foramen magnum and has a relatively high occurrence rate with cases of spina bifida with meningomyelocele.

Items 418-421

A 48 year-old man of Norwegian extraction complains of 'pins and needles' tingling sensations in all four extremities. He also states that on occasion he feels shooting pains in the arms and legs and a pressure sensation in the abdomen. The man walks with an unsteady, uncoordinated gait. On testing it is found that there is loss of vibratory and position sense in both lower extremities. Both pain and temperature sensation appear to be normal. There is noticeable muscular weakness in both lower limbs. Deep tendon reflexes are absent in the lower limbs but appear normal for the upper limbs. A bilateral Babinski sign can be elicited and the patient has a positive Romberg test. Other laboratory data reveal that the patient has primary anemia (vitamin B$_{12}$ deficiency).

418. Which of the following diagrams of spinal cord cross sections would most likely represent the extent of damage in this patient

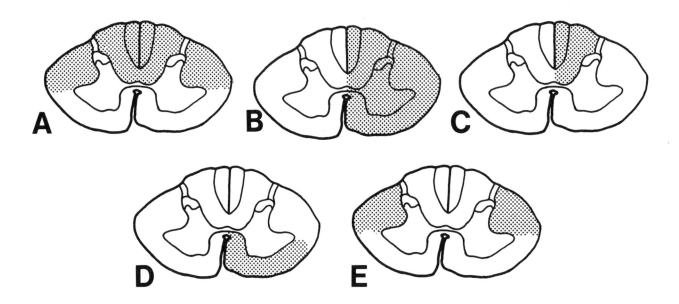

419. A patient with Babinski sign would exhibit

    (A) a tendency to sway and fall when asked to stand with the feet together and the eyes closed
    (B) a dorsiflexion of the large toe and a fanning of the lateral four toes
    (C) inability to touch the tip of the nose with the index finger
    (D) inability to correctly determine a figure such as a numeral drawn on the skin by the examiner
    (E) loss of pain and temperature sensation from the affected part of the body

420. The loss of vibratory and position sense is indicative of damage to the

    (A) spinothalamic tracts
    (B) spinocerebellar tracts
    (C) posterior columns
    (D) corticospinal tracts
    (E) rubrospinal tracts

421. The patient's description of shooting pains and the 'pins and needles' sensations are indicative of damage to the

   (A) spinothalamic tracts
   (B) corticospinal tracts
   (C) rubrospinal tracts
   (D) posterior roots
   (E) spinocerebellar tracts

ANSWERS AND TUTORIAL ON ITEMS 418-421

The answers are: **418-A; 419-B; 420-C; 421-D.** The neurological manifestations of pernicious (primary) anemia are degenerative changes in the peripheral nerves and central nervous system. The white matter of the brain appears to be more affected than the gray matter. Peripheral nerves, tracts in the spinal cord and the white matter of the brain all can show varying degrees of degeneration. The constellation of symptoms shown by this patient is termed 'combined system disease'. The characteristic early clinical symptoms are paresthesias in the distal parts of the extremities and a spastic, ataxic weakness of the legs. The presence of the bilateral Babinski signs in this patient indicates damage to the corticospinal tracts. Normally, when the sole of the foot is stroked firmly, the toes flex (plantar response). However, with damage to the corticospinal tracts, the response of the foot to stroking is extension of the big toe and a fanning of the lateral four toes (Babinski sign). The loss of vibratory and position sense indicates damage to the posterior columns. The 'pins and needles' sensations (paresthesias) indicate damage to the peripheral nerves or nerve roots.

Following occlusion of an artery, a patient has spinal cord damage represented by the shaded portion in the diagram below.

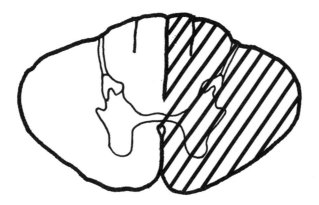

422. This spinal cord lesion is known clinically as

   (A) amyotrophic lateral sclerosis
   (B) Friedreich's ataxia
   (C) combined system disease
   (D) tabes dorsalis
   (E) Brown-Séquard syndrome

423. One type of sensory deficit shown by such a patient would include

   (A) loss of vibratory sense bilaterally in all four extremities
   (B) loss of vibratory sense in the lower extremity contralateral to the lesion
   (C) loss of vibratory sense in the lower extremity ipsilateral to the lesion
   (D) loss of vibratory sense in the upper extremity contralateral to the lesion
   (E) loss of vibratory sense bilaterally in the lower extremities

424. One type of motor deficit shown by such a patient would include

   (A) an ipsilateral upper motor neuron lesion below the level of the cord lesion
   (B) a contralateral upper motor neuron lesion below the level of the cord lesion
   (C) a lower motor neuron lesion in both lower extremities
   (D) an upper motor neuron lesion in both upper extremities
   (E) an upper motor neuron lesion in the contralateral upper extremity

425. One type of sensory deficit in such a patient would include

    (A) loss of pain and temperature sensation below and contralateral to the cord lesion
    (B) loss of pain and temperature sensation below and ipsilateral to the cord lesion
    (C) loss of pain and temperature sensation above and ipsilateral to the cord lesion
    (D) loss of pain and temperature sensation below and bilaterally
    (E) loss of pain and temperature sensation in both upper extremities

ANSWERS AND TUTORIAL ON ITEMS 422-425

The answers are: 422-E; 423-C; 424-A; 425-A. The signs and symptoms associated with hemisection of the cord constitute the Brown-Séquard syndrome. Neurological findings include: (1) loss of proprioceptive information carried by the posterior white columns from below the lesion on the same (ipsilateral) side. The posterior columns are the fasciculi gracilis and cuneatus which are comprised primarily of axon collaterals from type I and II dorsal root afferents. These axons carry proprioceptive information such as touch, pressure, vibration, and joint and position sense from the body, particularly from the limbs. (2) an upper motor neuron lesion below the level of the injury on the same side. Spinal hemisection destroys the fibers of the lateral corticospinal tract which originates in large part from the primary motor and sensory cortices of the contralateral cerebral hemisphere. These axons synapse on sensory interneurons and directly on motor neurons. Destruction of these fibers anywhere along their length constitutes an upper motor neuron lesion. (3) loss of pain and temperature sensation below the cord lesion on the opposite (contralateral) side. Spinal hemisection destroys the lateral and anterior spinothalamic tracts which carry pain, temperature and light touch information from the periphery. The axons synapse with interneurons in the substantia gelatinosa and dorsal horn. These axons of the second order neurons cross the midline via the anterior white commissure of the spinal cord, and thus a hemisection produces pain and temperature anesthesia contralateral to the lesion.

Items 426-429

A 62 year-old man presents with the classic symptoms of Parkinson's disease. The patient shows a moderate rhythmic tremor in both upper extremities. His face is mask-like (e.g. unexpressive) with few emotional movements. The normal involuntary blinking movements of the his eyes are decreased in frequency. When the patient walks, his head and shoulders are stooped forward and he slowly shuffles his feet. When walking, the patient keeps his arms somewhat extended and adducted and flexed. He does not swing his arms as he walks.

426. Disease or dysfunction of what structure of the brain has been implicated in Parkinson's disease

    (A) pulvinar
    (B) habenula
    (C) superior colliculus
    (D) substantia nigra
    (E) medial geniculate nucleus

427. The neural pathway implicated in Parkinson's disease is a dysfunction of the projection of the structure in question 426 above to the

    (A) cerebral cortex
    (B) spinal cord
    (C) cerebellum
    (D) superior colliculus
    (E) corpus striatum

428. Parkinson's disease is related to dysfunction in the production or release of the neurotransmitter

    (A) serotonin
    (B) dopamine
    (C) acetylcholine
    (D) norepinephrine
    (E) adrenalin

429. The current treatment for Parkinson's disease is to administer the amino acid precursor to the dysfunctional neurotransmitter. The neurotransmitter itself cannot be administered because it

    (A) is toxic to the patient
    (B) cannot be synthesized
    (C) does not cross the blood-brain barrier
    (D) would not have any effect on the disease process
    (E) acts too slowly to be of any benefit

ANSWERS AND TUTORIAL ON ITEMS 426-429

The answers are: **426-D; 427-E; 428-B; 429-C.** Parkinson's disease (paralysis agitans) is a complex of symptoms due to widespread, diffuse lesions in the basal ganglia and cerebral cortex and is characterized clinically by a variety of signs, which include a mask-like facial expression, dysarthria, alternating tremor, stooped posture, abnormalities of gait, cogwheel rigidity of the muscles, slowness and poverty of movements, lack of associated movements, disturbances of postural control and symptoms of autonomic nervous system dysfunction. These signs, along with the absence of evidence of pyramidal tract involvement, make a diagnosis conclusive. Parkinson's patients have been shown to have a reduction in the amount of dopamine in the corpus striatum and in the substantia nigra. The dopaminergic neurons of the substantia nigra project to the corpus striatum. L-dopa, the amino acid precursor to dopamine, is administered to patients because dopamine will not cross the blood-brain-barrier.

A 42 year-old woman presents with a sudden hemiplegia and hemianesthesia involving only the right leg. With testing it is found her symptoms also include mental confusion, some clouding of consciousness and aphasia.

430. Which artery (or major branch) has probably been occluded in this patient?

(A) Left anterior cerebral
(B) Right middle cerebral
(C) Right posterior cerebral
(D) Right anterior cerebral
(E) Left posterior cerebral

431. Concerning cerebral dominance, what can you likely conclude about this patient?

(A) She is right hemisphere dominant
(B) Both hemispheres have equal dominance
(C) Neither hemisphere is dominant
(D) Her left hemisphere is dominant
(E) No conclusion can be reached

432. If another patient had the opposite artery occluded than the one in the example above, which of the following symptoms probably would NOT be seen?

(A) Hemiplegia only
(B) Aphasia only
(C) Hemianesthesia only
(D) Hemiplegia or hemianesthesia
(E) Hemiplegia or aphasia

ANSWERS AND TUTORIAL ON ITEMS 430-432

The answers are: 430-A; 431-D; 432-B. The anterior cerebral artery supplies the anterior limb of the internal capsule, the head of the caudate nucleus and the putamen. Its distribution also includes the motor and sensory cortex which controls the legs. Since the corticospinal system decussates in the medulla, occlusion of the left anterior cerebral artery would produce the hemiplegia and hemianesthesia in the right leg. Since the anterior cerebral artery also supplies the white matter deep to Broca's area, occlusion of the left artery would also result in the aphasia seen in this patient. Most people (95%) are left cerebral hemisphere dominant, which means the language area, Broca's area, resides in that hemisphere. Thus a patient with occlusion of the right anterior cerebral artery is much less likely to exhibit aphasia.

Items 433-436

A CT scan of a patient reveals damage to the medulla oblongata in the region supplied by the posterior inferior cerebellar artery. The neurologist concludes the patient has a classic Wallenberg's syndrome (lateral medullary syndrome).

433. The patient would show an ipsilateral Horner's syndrome due to

    (A) damage to the mammillary bodies
    (B) destruction of descending sympathetic fibers
    (C) damage to the inferior olive
    (D) damage to the superior cervical ganglion
    (E) damage to the nucleus ambiguus

434. The patient would show dysphagia and dysarthria due to

    (A) destruction of the vestibular nuclei
    (B) destruction of the inferior olive
    (C) destruction of the nucleus ambiguus
    (D) damage to the medial lemniscus
    (E) damage to the pyramids

435. The patient would show a nystagmus due to

    (A) damage to the vestibular nuclei
    (B) damage to the nucleus ambiguus
    (C) damage to the inferior olive
    (D) destruction of the superior cerebellar peduncle
    (E) destruction of the medial lemniscus

436. The patient would have destruction of the spinal trigeminal tract and nucleus. This damage would produce

    (A) cerebellar type dysfunction in the ipsilateral arm and leg
    (B) impairment of pain and temperature sense in the ipsilateral leg
    (C) impairment of pain and temperature sense on the contralateral side of the face
    (D) impairment of taste sensation on the contralateral side of the face
    (E) impairment of pain and temperature sense on the ipsilateral side of the face

impaired speech

The answers are: **433-B; 434-C; 435-A; 436-E.** The posterior inferior cerebellar artery, which arises from the vertebral artery, supplies the lateral area of the medulla oblongata. The signs and symptoms that result from occlusion of this artery are called the lateral medullary syndrome or Wallenberg's syndrome. These signs and symptoms would include: dysphagia and dysarthria due to weakness of the ipsilateral palatal and laryngeal muscles (innervated by fibers originating in the nucleus ambiguus); impairment of pain and temperature sensation on the contralateral side of the body due to destruction of the ascending spinothalamic tracts; impairment of pain and temperature sense on the ipsilateral side of the face due to destruction of the spinal trigeminal nucleus and tract; nystagmus due to damage to the vestibular nuclei; cerebellar dysfunction in the ipsilateral arm and leg due to damage to the inferior cerebellar peduncle and the cerebellum. The patient would show ipsilateral Horner's syndrome due to destruction of the descending sympathetic fibers. The symptoms of Horner's syndrome would include decreased sweating on the face, flushed face due to vasodilation, and ptosis of the upper eyelid.

Items 437-441

A 53 year-old female patient has evidence of both a 'lower motor neuron lesion' and an 'upper motor neuron lesion' due to an expanding tumor in the spinal cord.

437. Characteristics of an upper motor neuron lesion include all of the following **EXCEPT**:

    (A) Babinski sign
    (B) exaggerated deep tendon reflexes
    (C) hypertonia
    (D) spastic paralysis
    (E) astereognosis

438. Characteristics of a lower motor neuron lesion include all of the following **EXCEPT**:

    (A) Babinski sign
    (B) flaccid paralysis
    (C) hypotonia
    (D) hyporeflexia
    (E) muscle atrophy

439. 'Upper motor neuron lesions' involving the spinal cord can be produced by damage to the

    (A) spinothalamic tract
    (B) posterior columns
    (C) corticospinal tract
    (D) spinocerebellar tracts
    (E) Lissauer's tract

440. 'Lower motor neuron lesions' are produced by damage to the

   (A) corticospinal tract
   (B) rubrospinal tract
   (C) ventral horn motor neurons
   (D) dorsal root ganglia
   (E) spinocerebellar tracts

441. In addition to injury to the spinal cord as in the example above, damage to which of the following structures would also result in a 'lower motor neuron lesion'?

   (A) Caudate nucleus
   (B) Substantia nigra
   (C) Red nucleus
   (D) Vestibular nuclei
   (E) Nucleus of the facial nerve

ANSWERS AND TUTORIAL ON ITEMS 437-441

The answers are: 437-E; 438-A; 439-C; 440-C; 441-E. A lower motor neuron lesion is defined as damage to the 'final common pathway', i.e., the motor neurons that synapse directly with skeletal muscle. In the spinal cord, lower motor neurons are the ventral horn motor neurons, but lower motor neurons also include neurons of cranial nerve nuclei (such the facial nerve nucleus) that innervate muscles of the head and neck. Damage to lower motor neurons causes a flaccid paralysis or paresis, decreased muscle tone (hypotonia), decreased reflexes (hyporeflexia) and eventual wasting or atrophy of the muscle. Upper motor neurons are defined as those neurons and axons of the corticospinal, corticopontine and corticobulbar tracts. A lesion to the 'upper motor neurons' that influence the spinal cord will cause a spastic paralysis or paresis, hyper-reflexia of the deep tendon reflexes, reduced or absent abdominal reflexes, hypertonia, Babinski sign and the clasp-knife reaction.

Items 442-446

    (A)   Diencephalon
    (B)   Midbrain
    (C)   Pons
    (D)   Upper medulla
    (E)   Lower medulla

Match the brain region listed above with the most appropriate description of that region or structures found within the region in the numbered questions below.

442. Location of the major structure involved in reflexive orientation of the eyes and the head to visual stimuli.

443. Location of the decussation of the corticospinal tract.

444. Location of the second order neurons that receive direct synapses from axons of the posterior column of the spinal cord.

445. Location of the motor nucleus for the muscles of facial expression.

446. Location of the portion of the ventricular system called the cerebral aqueduct.

ANSWERS AND TUTORIAL ON ITEMS 442-446

The answers are: **442-B; 443-E; 444-E; 445-C; 446-B**. The superior colliculus, which receives direct visual input from the optic tracts and is responsible for orientation of the eyes and head to visual stimuli. The superior colliculi are located in the tectum of the midbrain and are the larger of the four swellings called the corpora quadrigemina. The portion of the ventricular system found in the midbrain is the cerebral aqueduct. The tectum forms the roof for the cerebral aqueduct while the tegmentum of the midbrain forms the floor. The nucleus of the facial nerve (cranial nerve VII) which provides axons that innervate the muscles of facial expression, is found at the level of the pons. Both the dorsal (posterior) column nuclei and the pyramidal decussation are found at the level of the caudal end (lower) of the medulla. The dorsal column nuclei includes the nucleus gracilis which receives axons from the fasciculus gracilis of the spinal cord and the nucleus cuneatus which receives axons from the fasciculus cuneatus of the spinal cord. The dorsal column nuclei constitute the second order neurons which convey proprioceptive information from the extremities to the thalamus. The corticospinal tracts originate from the primary motor and sensory cortices of the cerebrum and pass through the entire brainstem on their way to the spinal cord. Approximately 85-90% of the axons of the each corticospinal tract decussate within the caudal medulla to continue as the lateral corticospinal tract of the cord. The other 10-15% remain ipsilateral as the anterior corticospinal tract.

Items 447-450

A 57 year-old man presents with paralysis and atrophy of the ipsilateral half of the tongue, paralysis of the contralateral arm and leg and impairment of tactile and position sensation in the trunk and extremities on the paralyzed side.

447. The site of the lesion that would produce these symptoms would likely be the

    (A) spinal cord
    (B) lateral portions of midbrain
    (C) medial region of the medulla oblongata
    (D) lateral portions of the pons
    (E) midline of the cerebellum

448. The paralysis of the contralateral arm and leg is due to damage to the

    (A) spinothalamic tracts
    (B) medial lemniscus
    (C) inferior olive
    (D) middle cerebellar peduncle
    (E) pyramidal tracts

449. The impairment of tactile and position sense contralateral to the lesion is due to damage to the

    (A) pyramidal tracts
    (B) inferior olive
    (C) middle cerebellar peduncle
    (D) medial lemniscus
    (E) nucleus ambiguus

450. The paralysis and atrophy of the tongue is due to damage to the

    (A) nucleus ambiguus
    (B) spinal trigeminal nucleus
    (C) facial nucleus
    (D) hypoglossal nucleus
    (E) medial lemniscus

ANSWERS AND TUTORIAL ON ITEMS 447-450

The answers are: **447-C; 448-E; 449-D; 450-D.** This patient exhibits a vascular lesion involving the medial portions of the medulla (medial medullary syndrome). The paramedian area of the medulla is nourished by perforating branches which arise from the vertebral arteries. Occlusion of one of these vessels would produce a lesion involving the pyramids and since the pyramids contain the descending corticospinal tracts, such a lesion would give rise to an upper motor neuron deficit of the contralateral extremities. Involvement of the medial lemniscus would result in loss of proprioceptive information (tactile and position sense) from the contralateral portion of the body. The medial lemnisci are comprised of axons originating in the dorsal column nuclei (nucleus cuneatus and gracilis). Involvement of the hypoglossal nucleus and/or the hypoglossal nerve fibers exiting the brain stem would give rise to the paralysis and atrophy of the ipsilateral half of the tongue. The hypoglossal nucleus lies in the dorsal aspect of the medulla just deep to the fourth ventricle. The axons descend through the medulla to exit the brainstem between the inferior olivary nucleus and the pyramids.

Items 451-454

A 33 year-old woman complains of double vision and a drooping of her upper eyelids. She states that, initially, the visual problems and drooping were transient, but they are now constant. When she speaks, it is noted that her voice is nasal and weak, and with prolonged talking, her voice fatigues to the point of unintelligibility. She states that her throat becomes tired when she eats a large meal. Her face shows little movement and when she attempts to smile, her expression resembles a snarl more than a smile. As she speaks, she holds a hand under her jaw. Blood tests reveal the patient has a abnormally high serum titer of antibodies to human acetylcholine receptors.

451. The diagnosis for this patient is likely to be

    (A) Raynaud's disease
    (B) Parkinson's disease
    (C) Weber's syndrome
    (D) myasthenia gravis
    (E) Reye's syndrome

452. The symptoms shown by this patient are the consequence of a

    (A) loss of acetylcholine receptors at the neuromuscular junction
    (B) lesion within the caudate nucleus
    (C) lesion within the red nucleus and substantia nigra
    (D) loss of skeletal muscle fibers
    (E) loss of ventral horn motor neurons

453. Innervation of the skeletal muscle that raises the upper eyelid is by cranial nerve

    (A) III
    (B) VII
    (C) V
    (D) IV
    (E) VI

454. The clinician orders a CT scan of the mediastinum for this patient. The reason for this is because

  (A) there may be an aneurysm of the aortic arch
  (B) the patient may have left ventricular hypertrophy
  (C) the thymus gland may have some pathogenetic role in this patient's disease process
  (D) there may be an esophageal stenosis or atresia
  (E) there may be a congenital heart defect

ANSWERS AND TUTORIAL ON ITEMS 451-454

The answers are: **451-D; 452-A; 453-A; 454-C**. This patient suffers from myasthenia gravis which is characterized by weakness and undue fatigability with exercise or muscular use. It most frequently affects the oculomotor, facial, laryngeal, pharyngeal, proximal limb and respiratory muscles. The facial muscle involvement gives rise to a characteristic smile call the 'myasthenic snarl'. Myasthenia gravis is an autoimmune syndrome that results in the loss of acetylcholine receptors from the postsynaptic membrane of the neuromuscular junction. It has been demonstrated that 85 percent of patients have circulating antibodies against human acetylcholine receptors. The thymus gland appears to play an as yet unidentified pathogenetic role in myasthenia gravis. While a CT scan of the mediastinum is not a diagnostic test, 65 percent of patients with this disease have thymic hyperplasia and 10 percent have thymomas.

Items 455-459

A 17 year-old boy was working as a golf caddie when he was struck by a golf ball on the right side of the temple just above and in front of the ear. The boy was transported unconscious to the hospital. In the hospital emergency room, the boy regained consciousness for a brief period, but then lapsed into unconsciousness again. On examination, the right pupil was dilated and there was decreased muscle tone in the left leg. A Babinski reflex was elicited from the left leg but not from the right. Lumbar puncture revealed an increased CSF pressure and some blood in the CSF. A skull radiograph revealed a compressed fracture of the anterior, inferior angle of the parietal bone.

455. A likely diagnosis of this patient is

  (A) subdural hematoma
  (B) contusion of the cervical spinal cord
  (C) extradural hemorrhage
  (D) Arnold-Chiari malformation
  (E) occlusion of the cerebral aqueduct

456. If your diagnosis includes rupture or tearing of a blood vessel, which vessel or one of its branches is most likely damaged in this patient?

  (A) Middle meningeal artery
  (B) Middle cerebral artery
  (C) Superior sagittal sinus
  (D) Vertebral artery
  (E) Internal carotid artery

457. What portion of the brain is injured to account for the hemiplegia and Babinski sign on the left side?

(A) Left medulla oblongata
(B) Right occipital lobe
(C) Right precentral gyrus
(D) Left temporal lobe
(E) Right pons and cerebellum

458. Damage to which cranial nerve would explain the dilated pupil?

(A) Right trochlear nerve
(B) Left trigeminal nerve
(C) Left optic nerve
(D) Right oculomotor nerve
(E) Right optic nerve

459. What test would you employ to detect possible blockage of the subarachnoid space?

(A) Weber's test
(B) Romberg's test
(C) Allen's test
(D) Tinel's test
(E) Queckenstedt's test

ANSWERS AND TUTORIAL ON ITEMS 455-459

The answers are: **455-C; 456-A; 457-C; 458-D; 459-E.** This patient has a right-sided extradural hemorrhage due to fracture of the parietal bone in the region of the pterion. The anterior division of the middle meningeal artery (and vein) passes directly beneath this portion of the skull and is in danger of tearing with a fracture. The initial trauma is responsible for the patient being found unconscious, but recovery of consciousness for a period only to relapse into unconsciousness is characteristic of such injuries. The relapse is due to the accumulation of a large blood clot (hematoma) between the skull and the dura mater. The hematoma places pressure on the right precentral gyrus of the cerebrum and causes the hemiplegia and Babinski sign on the left side. The expanding blood clot probably places indirect pressure on the right oculomotor nerve which is responsible for the dilated pupil on the right side. The clot is also responsible for the raised cerebrospinal fluid pressure. Queckenstedt's test (compressing the internal jugular vein and observing CSF pressure changes) is used to detect blockage of the subarachnoid space. Blood in the CSF is probably due to a small leakage from the extradural space into the subarachnoid space at the fracture site.

The diagram below represents the visual pathways. The labeled bars represent regions where damage has occurred to the visual pathways. Use this diagram to answer the questions that follow.

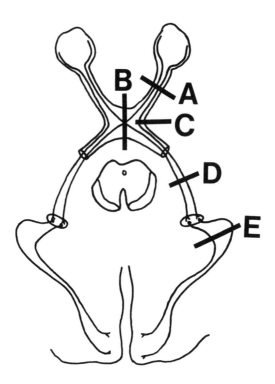

460. Which of the lesions above would produce a right nasal hemianopia?

    (A) A
    (B) B
    (C) C
    (D) D
    (E) E

461. Which of the lesions above would produce a bitemporal heteronymous hemianopia?

    (A) A
    (B) B
    (C) C
    (D) D
    (E) E

462. Which of the lesions above includes the portion of the optic radiations called Meyer's loop?

    (A) A
    (B) B
    (C) C
    (D) D
    (E) E

463. Which of the following representations of the visual fields corresponds to a lesion at E.

464. Which of the following representations of the visual fields corresponds to a lesion at A.

ANSWERS AND TUTORIAL ON ITEMS 460-464

The answers are: **460-C; 461-B; 462-E; 463-C; 464-A**. The diagram represents the visual pathways. Axons within the optic nerves partially decussate at the optic chiasm with axons originating from the temporal retina remaining ipsilateral while those originating from the nasal retina cross at the chiasm to synapse in the contralateral lateral geniculate nucleus (LGN). From the LGN axons, travel as the optic radiations to synapse mainly within the primary visual cortex (Area 17 of Brodmann). A complete lesion of the optic nerve (point A) would produce a total blindness in the affected eye. A lesion that involves only the uncrossed fibers of the chiasm (point C) would produce a right nasal hemianopia. A lesion that involved only the decussating fibers of the optic chiasm (point B) would produce a bitemporal heteronymous hemianopia. A portion of the optic radiations called Meyer's loop (point E) extends into the temporal lobe and carries axons representing the superior visual quadrants.

The diagram below represents a coronal section through the brain.

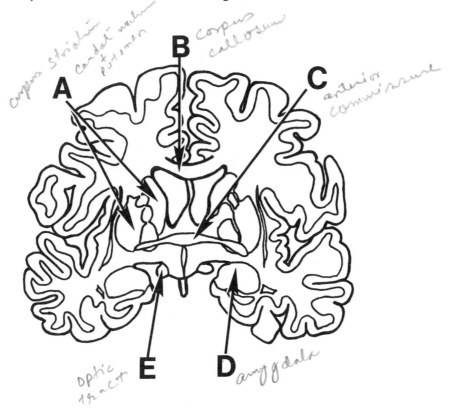

465. Structure which represents the amygdala.

    (A) A
    (B) B
    (C) C
    (D) D
    (E) E

466. Fiber bundle that interconnects the rostral portions of the temporal lobes.

    (A) A
    (B) B
    (C) C
    (D) D
    (E) E

467. Fiber bundle that contains efferents of retinal ganglion neurons.

    (A) A
    (B) B
    (C) C
    (D) D
    (E) E

468. The structure labeled at B represents a large fiber bundle called the

   (A) internal capsule
   (B) anterior commissure
   (C) fornix
   (D) corpus callosum
   (E) pyramidal decussation

469. The structures labeled at A are sometimes referred to collectively as the

   (A) corpus striatum
   (B) insula
   (C) island of Reil
   (D) corpora quadrigemina
   (E) trigone

ANSWERS AND TUTORIAL ON ITEMS 465-469

The answers are: **465-D; 466-C; 467-E; 468-D; 469-A**. This diagram represents a coronal section of the brain through the level of the anterior commissure. Structures labeled are: A - head of caudate nucleus and putamen (corpus striatum); B - corpus callosum; C - anterior commissure; D - amygdala; E - optic tract. The head of the caudate nucleus and the putamen/globus pallidus complex are frequently referred to as the corpus striatum. The anterior commissure is composed of commissural fibers interconnecting the rostral portions of the temporal lobes. The optic tracts contain efferent axons from the retinal ganglion cells. The anterior limb of the internal capsule lies at the level of the anterior commissure and divides the head of the caudate nucleus from the lenticular nucleus. The amygdala is a large nucleus that lies in the rostral pole of the temporal lobe and is considered part of the limbic system.

Items 470-473

A 39 year-old woman was seen by the neurologist after fainting and remaining unconscious for several hours. When consciousness returned, she appeared confused and was unable to speak. Examination revealed a spastic paralysis of her right arm but no atrophy. Her right leg and both left extremities appeared normal. Her tongue protruded to the right but was not atrophic. Her right facial muscles were paralyzed, but only those below the eye. There were no apparent deficits in pain or temperature or other somesthetic modalities. There appeared to be no visual defects. The medical history of this patient revealed she had suffered from subacute bacterial endocarditis about a year and a half prior to the present situation.

470. Given these signs and symptoms, the likely diagnosis for this patient is

   (A) embolism at the origin of the right internal carotid artery
   (B) embolism in the basilar artery
   (C) embolism in the superior sagittal sinus
   (D) embolism in a branch of the left middle cerebral artery
   (E) embolism in the right middle cerebral artery

471. Which of the following diagrams would most accurately depict the extent of the lesion in this patient?

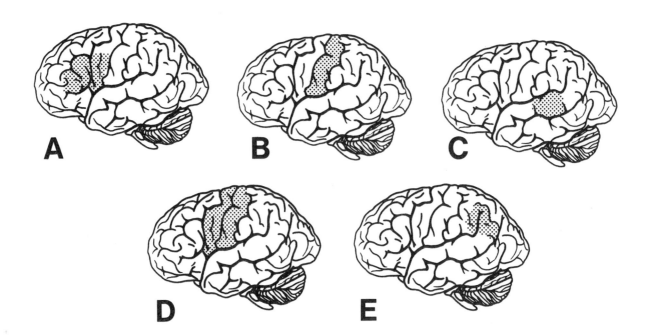

472. Which of the following is the most likely explanation for the lower facial paralysis in this patient?

(A) The portion of the facial nucleus controlling the lower face receives a bilateral cortical projection while the portion controlling the upper face receives only a contralateral cortical projection.
(B) The entire facial nucleus receives only a contralateral cortical projection.
(C) The entire facial nucleus receives a bilateral cortical projection.
(D) The entire facial nucleus receives only an ipsilateral cortical projection.
(E) The portion of the facial nucleus controlling the upper face receives a bilateral cortical projection while the portion controlling the lower face receives only a contralateral cortical projection.

473. The type of speech deficit seen in this patient is called

(A) alexia without agraphia
(B) Wernicke's aphasia
(C) Broca's aphasia
(D) alexia with agraphia
(E) conduction aphasia

ANSWERS AND TUTORIAL ON ITEMS 470-473

The answers are: 470-D; 471-D; 472-E; 473-C. The spastic paralysis of only the right upper extremity suggests damage to a restricted portion of the cerebral cortex. A lesion of the subcortical white matter, internal capsule or cerebral peduncle would most likely have resulted in paralysis of the lower extremity

206

as well. A lesion within the internal capsule would also have produced sensory deficits not seen in this patient. Upper motor neuron deficits of both the tongue and the face further support this contention. The portion of the facial nucleus that controls the upper facial muscles receives a bilateral projection from the motor cortex while the portion of the nucleus controlling the lower facial muscles receives a contralateral projection only. Thus, with cortical damage, facial paralysis is evident only in those muscles below the eye. The presence of Broca's (motor) aphasia further localizes the damage to the frontal opercular area of the dominant (usually left) hemisphere. An embolism in a branch of the left middle cerebral artery would produce the signs and symptoms seen in this patient. Emboli in the arterial supply to the brain are common sequelae in patients with a history of bacterial endocarditis.

Items 474-479

The diagram below is a lateral view of the brain.

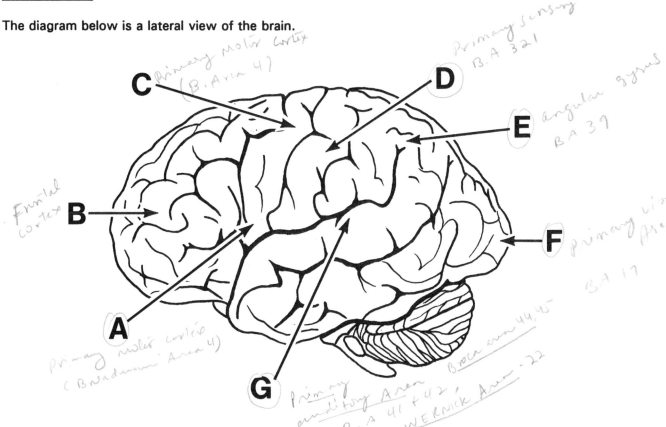

474. Region of cerebral hemisphere where damage would produce the loss of ability to read (alexia) and write (agraphia).

475. Region of cerebral hemisphere that receives direct connections from the lateral geniculate nucleus of the thalamus.

476. Region of the cerebral hemisphere where damage would result in Wernicke's aphasia.

477. Region of the cerebral hemisphere that would control volitional movement of the hand and fingers.

478. Region of the cerebral hemisphere known as the primary somesthetic area or Brodmann's areas 3, 1 and 2.

479. Region of the cerebral hemisphere known as the primary auditory cortex.

The answers are: **474-E; 475-F; 476-G; 477-A; 478-D; 479-G**. The diagram shows a lateral view of the left hemisphere of the brain. Labeled areas are: A - upper extremity portion of the precentral gyrus (primary motor cortex, Brodmann's area 4); B - frontal cortex; C - precentral gyrus (primary motor cortex, Brodmann's areas 4); D - primary sensory cortex (Brodmann's areas 3, 2 and 1); E - angular gyrus (Brodmann's area 39); F - primary visual area (Brodmann's area 17); G - primary auditory area (Brodmann's area 41 and 42, Wernicke's area). The angular gyrus is the caudal part of the inferior parietal lobule and is interconnected with the supramarginal gyrus and Wernicke's area. It also receives connections from the areas 17, 18 and 19. A lesion of this area in the dominant hemisphere results in the loss of ability to read (alexia) and write (agraphia). The primary visual cortex (area 17) receives direct connections from the lateral geniculate nucleus of the thalamus. A lesion of the primary auditory cortex (areas 41 and 42, Wernicke's area) in the dominant hemisphere causes a fluent, para-grammatical aphasia (also known as auditory receptive or sensory aphasia). The precentral gyrus which contains the primary motor cortex is responsible for initiation of voluntary movement of skeletal muscles of the limbs. The postcentral gyrus (primary somesthetic area) receives pain, temperature and other somesthetic information relayed from certain thalamic nuclei.

Items 480-483

At autopsy, the pathologist explains that the patient, a 5 year-old child, had a medulloblastoma that appeared to have arisen in the vermis of the cerebellum and had invaded the 4th ventricle and the neighboring cerebellar hemispheres.

480. When alive, this child would have shown all of the following signs and symptoms **EXCEPT**:

    (A) astereognosis
    (B) ataxia
    (C) intention tremor
    (D) nystagmus
    (E) wide-based stance

481. Other signs or symptoms associated with cerebellar disease would include all of the following **EXCEPT**:

    (A) dysdiadochokinesia
    (B) decomposition of movement
    (C) scanning speech
    (D) hypotonia
    (E) paralysis

482. The cerebellum probably receives input from all of the following structures or areas **EXCEPT**:

    (A) vestibular nuclei
    (B) hippocampus
    (C) pontine nuclei
    (D) motor and premotor cortex
    (E) spinal cord

483. Efferents from the cerebellum project directly to all of the following EXCEPT:

    (A) dentate nuclei
    (B) lateral geniculate nuclei
    (C) globose nuclei
    (D) emboliform nuclei
    (E) fastigial nuclei

ANSWERS AND TUTORIAL ON ITEMS 480-483

The answers are: **480-A; 481-E; 482-B; 483-B**. Disorders of the cerebellum result in distinctive symptoms and signs and can often be localized to specific portions of the cerebellum. The most common lesion involving the vestibulocerebellum (vermis and flocculonodular lobe) is the medulloblastoma, a rapidly growing tumor usually occurring in childhood and usually fatal within a year. Involvement of the fourth ventricle would result in increased intracranial pressure and internal hydrocephalus. Patients with cerebellar disease would not show loss of stereognosis since this information is carried to the thalamus via the medial lemniscal system. Patients with cerebellar disease have considerable loss of muscular coordination but would not show paralysis. The hippocampus is part of the limbic system and probably has no connections with the cerebellum. The Purkinje cells of the cerebellum project directly to the four deep cerebellar nuclei - the dentate, globose, emboliform and fastigial nuclei.

A 45 year-old woman states that she has suffered from headaches for several months. She now complains of double vision. Lately she has noticed that her left arm seems weak and she has become increasingly more clumsy with her left hand (e.g. dropping dishes and other small items). On examination, her left ankle and knee deep tendon reflexes were exaggerated and a Babinski sign was elicited on the left. An eye examination revealed that her right pupil was larger than the left and her right eye was turned outward and downward. When asked to look at an object placed close to her nose, her right eye did not converge and neither a light nor an accommodation reflex could be elicited from the right eye. The muscles of the right side of her face below the eye were paralyzed and when asked to protrude her tongue, it deviated to the right.

484. A CT scan showed this patient to have a tumor. Given these signs and symptoms, where would you expect this tumor to lie?

    (A) Caudal medulla at the level of the pyramidal decussation
    (B) Rostral medulla at the level of the obex
    (C) At the level of the cerebral peduncle near the emergence of the right oculomotor nerve.
    (D) Cerebellopontine angle
    (E) Sella turcica

485. The origin of the axons which control pupillary constriction with light stimulation and during accommodation are found in the

    (A) lateral geniculate nucleus
    (B) superior colliculus
    (C) nucleus of the trochlear nerve
    (D) nucleus solitarius
    (E) nucleus of Edinger-Westphal

486. The facial paralysis and tongue paralysis in this patient are likely due to damage to the

    (A) hypoglossal nerve directly
    (B) facial nerve directly
    (C) corticobulbar fibers to the nuclei of the facial and hypoglossal nerves
    (D) medial lemniscus
    (E) medial longitudinal fasciculus

487. Preganglionic parasympathetic nerve fibers carried by the oculomotor nerve synapse in the

    (A) ciliary ganglion
    (B) otic ganglion
    (C) superior cervical ganglion
    (D) pterygopalatine ganglion
    (E) nodose ganglion

ANSWERS AND TUTORIAL ON ITEMS 484-487

The answers are: **484-C; 485-E; 486-C; 487-A**. This patient exhibits the signs and symptoms of hemiplegia alternata oculomotoria, or Weber's syndrome. This can be caused by a vascular lesion to the midbrain or by a tumor adjacent to the cerebral peduncle and the emergence of the oculomotor nerve (cranial nerve III). The facial paralysis and tongue paralysis are both due to damage to the descending corticobulbar fibers that innervate the nuclei of the facial and hypoglossal nerves. Damage to the oculomotor nerve produces ptosis of the eyelid and paralysis of all the extrinsic muscles of the eye except the superior oblique and the lateral rectus (which explains the position of the eye in this patient). The oculomotor nerve also carries the preganglionic parasympathetic nerve fibers which originate in the nucleus of Edinger-Westphal and synapse in the ciliary ganglion. The postganglionic fibers innervate the ciliaris muscles and the sphincter pupillae. Damage to these fibers results in pupillary dilation and the lack of response of the pupil to light or accommodation.

Items 488-490

A 41 year-old homeless man was admitted to the emergency room exhibiting confusion, disorientation, amnesia and confabulation. During a lucid period, the man admitted that he drank a large amount of cheap wine and he had done so for years. Alcohol was found in the patient's blood and also in the CSF following a spinal tap. The patient showed a paralysis of both lateral recti, horizontal and vertical nystagmus and a paralysis of conjugate gaze. He also had weakness of the legs, foot drop and ataxia in walking. The muscular weakness appeared to be greatest in the distal parts of the lower limbs. Reflexes were absent at the ankle and knee, the plantar response was absent and the abdominal skin reflexes were decreased. There was anesthesia in the distal part of his lower limb and vibratory and kinesthetic sensibilities were impaired. The nerves of his lower limbs were sensitive to pressure and the muscles were painful when squeezed.

488. The peripheral nerve disturbances in this patient are classified as

    (A) polyneuritis
    (B) tabes dorsalis
    (C) combined system disease
    (D) myasthenia gravis
    (E) pseudobulbar palsy

489. The mental symptoms shown by this patient are classified as

    (A) Froehlich's syndrome
    (B) amyotrophic lateral sclerosis
    (C) Paget's disease
    (D) Hurler's syndrome
    (E) Korsakoff's syndrome

490. The mental symptoms shown by this patient suggest lesions within the limbic system of the brain. The limbic system would include all of the following EXCEPT:

(A) hippocampus
(B) cerebellum
(C) cingulate gyrus
(D) mammillary bodies
(E) amygdaloid nuclei

ANSWERS AND TUTORIAL ON ITEMS 488-490

The answers are: **488-A; 489-E; 490-B**. Alcohol-vitamin deficiency polyneuritis is the most common form of neuritis and occurs in patients addicted to the use of large amounts of alcohol and who have associated nutritional deficiencies. The pathology of polyneuritis is due mainly to a non-inflammatory degeneration of the peripheral nerves producing the pains, paresthesias, weakness and sensory loss. Muscular weakness is usually greatest in the distal part of the extremities. The confusion, disorientation, amnesia and confabulation seen in this patient are referred to as Korsakoff's syndrome. Patients with Korsakoff's syndrome exhibit pathological changes in diencephalic structures that are part of the limbic system. Typically, they have damage to the mammillary bodies as well as the medial dorsal nucleus of the thalamus. The cerebellum is not classified as part of the limbic system.

The figure is a transverse image of normal brain obtained by CT scanning.

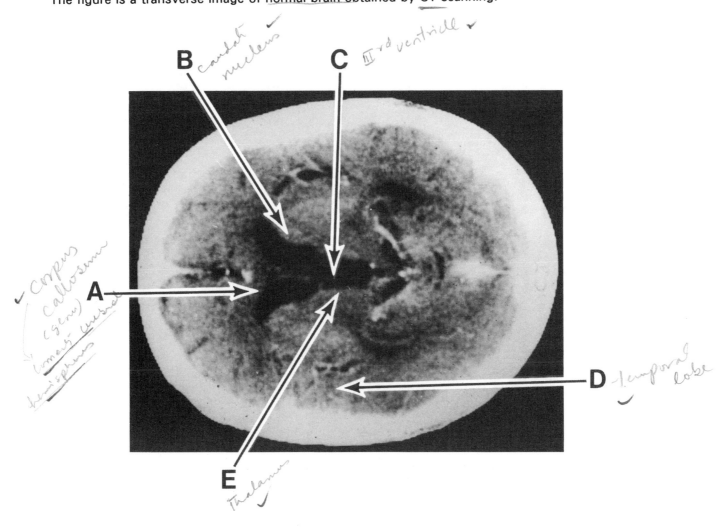

Handwritten annotations on figure:
- B — caudate nucleus ✓
- C — IIIrd ventricle ✓
- A — corpus callosum (genu) internal capsule hemispheres
- D — temporal lobe ✓
- E — Thalamus ✓

491. Region of the brain were a lesion such as epilepsy could profound effects on learning and memory.

492. Region of the brain which is divided into the rostrum, genu, body and splenium. (corpus callosum)

493. Portion of the ventricular system that extends into the infundibulum and tuber cinereum.

494. Region of the brain where the internal structures include the hippocampus and amygdala.

✓ 495. Region of the brain which is a part of the basal ganglia.   corpus striatum ① + ②

Handwritten note: Limbic system!

corpus striatum ① + ②
- ① caudate nucleus
- ② Lentiform Nucleus (Putamen + Globus pallidus)
- ③ claustrum
- ④ amygdala

The figure is a MRI scan of a midsagittal slice through the head.

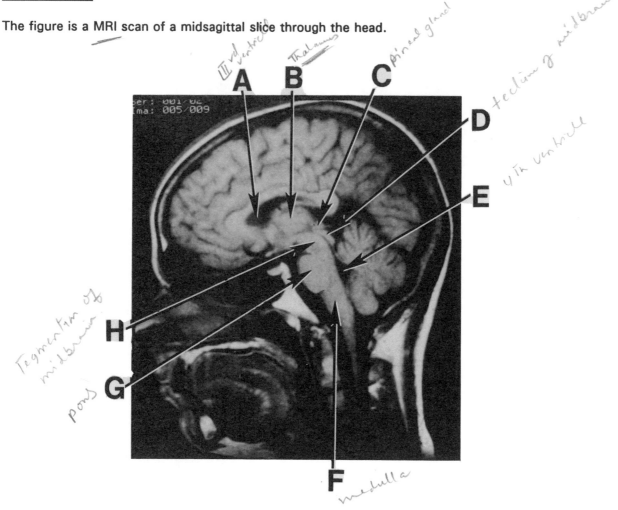

Handwritten annotations on figure: III rd ventricle (A), Thalamus (B), Pineal gland (C), tectum of midbrain (D), 4th ventricle (E), Tegmentum of midbrain (H), pons (G), medulla (F)

496. Region of the brain known as the corpora quadrigemina which function to control visual and auditory reflexes.

497. Portion of the ventricular system where CSF exits into the subarachnoid space via the foramen of Magendie and foramina of Luschka.

498. Site of tumor of the pineal gland which could compress aqueduct of Sylvius.

499. Region of the brain that is described as the major sensory relay nucleus of the brain.

500. Region of brain were a vascular accident could compromise the red nucleus and substantia nigra.

501. Region of the brain that contains the principle sensory nucleus of and gives origin to the fibers of the trigeminal nerve.

ANSWERS AND TUTORIAL ON ITEMS 491-495

The answers are: **491-D; 492-A; 493-C; 494-D; 495-B**. The figure is a transverse CT scan through the third ventricle and thalamus. Labeled structures are; A - corpus callosum; B - head of caudate nucleus; C - third ventricle; D - temporal lobe; E - thalamus. The corpus callosum is comprised of axons interconnecting the cerebral hemispheres and is divided into the rostrum, genu, body and splenium. The genu of the callosum is shown in A. The 3rd ventricle, which lies between the two halves of the thalamus, has an extension that descends into the infundibulum of the pituitary. The hippocampus and amygdala, both portions of the limbic system, are found within the temporal lobes of the cerebrum and lesions of this part of the brain leads to striking deficits in learning and memory. The head of the caudate nucleus is seen in this scan and it, plus the putamen and globus pallidus, comprise the major portion of the basal ganglia.

[handwritten: Corpus striatum = Putamen + Caudate Nucleus + Globus pallidus]
[handwritten: or Basal ganglia]

ANSWERS AND TUTORIAL ON ITEMS 496-501

The answers are: **496-D; 497-E; 498-C; 499-B; 500-H; 501-G**. The figure is a magnetic resonance image (T1 MRI) of a midsagittal section through the head. Labeled structures are: A - third ventricle; B - thalamus; C - pineal gland; D - tectum of midbrain; E - 4th ventricle; F - medulla; G - pons; H - tegmentum of midbrain. The roof of the cerebral aqueduct is known as the tectum and consists primarily of the corpora quadrigemina, the superior and inferior colliculi. Cerebrospinal fluid (CSF) circulates throughout the ventricular system of the brain. Three foramina in the roof of the 4th ventricle (foramen of Magendie and foramina of Luschka) allow the CSF to enter the subarachnoid space. A tumor of the pineal gland (C) may compress the anterior portion of the tectum and occlude the aqueduct of Sylvius causing an internal hydrocephalus. The thalamus (B) is the major sensory relay nucleus of the brain, receiving all sensory information except olfactory. The red nucleus and the substantia nigra are part of the extrapyramidal motor system and are large nuclei found within the tegmental region of the midbrain. The principle sensory nucleus of cranial nerve V (trigeminal nerve) lies within the pons and the fibers of the trigeminal nerve exit the brain stem from the lateral portion of the pons.

A 5½ year-old boy was brought to the emergency room by his parents. They said their son had complained earlier of a headache and a stiff neck. During the night he had a fever, had vomited and had a seizure. On examination, the neck stiffness persisted, the neck was tender to the touch and it was very difficult for the examiner to flex the neck muscles. On attempting to flex the head, there was a flexion of the legs at the knees. When the thigh was flexed, it was extremely difficult to extend the leg completely due to pain in the back. The boy seemed to be lethargic, delirious and almost in a stupor. The boy's temperature was 102° F. The boy showed no evidence of cranial nerve palsies or focal neurological signs.

502. Given these signs and symptoms, the likely diagnosis is

(A) epidural hematoma
(B) meningitis
(C) cerebrovascular accident
(D) intracerebral tumor
(E) epilepsy

503. In this patient, when the neck was passively flexed, flexion of the leg occurred. This is known as

(A) Allen's test
(B) Brudzinski's sign
(C) Tinel's sign
(D) Kernig's sign
(E) Trendelenburg's sign

504. The analysis of the CSF and CSF pressure of this patient would likely show all of the following EXCEPT:

(A) increased white blood cell content
(B) increased protein levels
(C) clear and colorless consistency to the CSF
(D) increased CSF pressure
(E) presence of microorganisms

505. Normally in these patients, CSF is obtained

(A) from the lateral ventricles
(B) from the cisterna magna
(C) from the central canal
(D) from the superior sagittal sinus
(E) by lumbar puncture

506. When this patient was on his back and his thigh was flexed on the trunk, extension of the leg was nearly impossible due to back pain. This is known as

    (A) Kernig's sign
    (B) Trendelenburg's sign
    (C) Tinel's sign
    (D) Brudzinski's sign
    (E) Allen's test

ANSWERS AND TUTORIAL ON ITEMS 502-506

The answers are: **502-B; 503-B; 504-C; 505-E; 506-A**. There are several causes of meningitis but the signs and symptoms are basically the same. The onset of meningitis is accompanied by chills and fever, headache, nausea and vomiting, pain in the back and stiffness of the neck. Mental confusion and stupor may occur and convulsive seizures are common. The temperature is usually elevated to 101° to 103° F. There is rigidity of the neck and the patient usually shows both a positive Kernig's and Brudzinski's sign. A lumbar puncture is done and the laboratory data show increased CSF pressure, and the fluid is cloudy or purulent and contains many leukocytes. The protein content is increased, the sugar content is decreased and organisms can be seen in stained smears of the fluid and can be cultured on the appropriate media.